TUNING
THE WORLD

NEW MATERIAL HISTORIES *of* MUSIC

A series edited by James Q. Davies *and* Nicholas Mathew

ALSO PUBLISHED IN THE SERIES:

Musical Vitalities: Ventures in a Biotic Aesthetics of Music
Holly Watkins

Sex, Death, and Minuets: Anna Magdalena Bach and Her Musical Notebooks
David Yearsley

The Voice as Something More: Essays toward Materiality
Edited by Martha Feldman and Judith T. Zeitlin

Listening to China: Sound and the Sino-Western Encounter, 1770–1839
Thomas Irvine

The Search for Medieval Music in Africa and Germany, 1891–1961: Scholars, Singers, Missionaries
Anna Maria Busse Berger

An Unnatural Attitude: Phenomenology in Weimar Musical Thought
Benjamin Steege

Mozart and the Mediation of Childhood
Adeline Mueller

Musical Migration and Imperial New York: Early Cold War Scenes
Brigid Cohen

The Haydn Economy: Music, Aesthetics, and Commerce in the Late Eighteenth Century
Nicholas Mathew

Tuning the World

THE RISE OF 440 HERTZ IN MUSIC, SCIENCE & POLITICS, 1859–1955

Fanny Gribenski

The University of Chicago Press
Chicago and London

The University of Chicago Press, Chicago 60637
The University of Chicago Press, Ltd., London

Published 2023
Printed in the United States of America

32 31 30 29 28 27 26 25 24 23   1 2 3 4 5

ISBN-13: 978-0-226-82326-3 (cloth)
ISBN-13: 978-0-226-82327-0 (e-book)
DOI: https://doi.org/10.7208/chicago/9780226823270.001.0001

This book has been supported by the Joseph Kerman Fund and General Fund of the American Musicological Society, supported in part by the National Endowment for the Humanities and the Andrew W. Mellon Foundation.

Library of Congress Cataloging-in-Publication Data

Names: Gribenski, Fanny, author.
Title: Tuning the world : the rise of 440 Hertz in music, science, and politics, 1859–1955 / Fanny Gribenski.
Other titles: Rise of 440 Hertz in music, science, and politics, 1859–1955 | New material histories of music.
Description: Chicago: The University of Chicago Press, 2023. | Series: New material histories of music | Includes bibliographical references and index.
Identifiers: LCCN 2022022556 | ISBN 9780226823263 (cloth) | ISBN 9780226823270 (ebook)
Subjects: LCSH: Tuning—History—19th century. | Tuning—History—20th century. | Musical pitch—Standards—History. | BISAC: MUSIC / History & Criticism | MUSIC / Philosophy & Social Aspects
Classification: LCC ML3809 .G66 2023 | DDC 781.2/32—dc23/eng/20220513
LC record available at https://lccn.loc.gov/2022022556

♾ This paper meets the requirements of ANSI/NISO Z39.48-1992 (Permanence of Paper).

To Ian and Zoë

Contents

List of Illustrations ix

Introduction: Tuning Forks and Global Politics 1

1. Tuning the Nation: Aesthetics, Science, Industry, and the French Pitch 27

2. Sounding the World: Nationalism, Internationalism, and the Travels of the French Pitch 59

3. Retuning the World: Transatlanticism and the Defeat of the French Pitch 99

4. "Pitch in Our Time": International Concord and the Engineering of an Interwar Standard 133

5. Postwar Aftermath: Confirming an Embattled Standard 161

Epilogue 187

Acknowledgments 195

Notes 199

Index 259

Illustrations

1.1 *Mémoire sur l'étude optique des mouvements vibratoires* (1857) · 34
1.2 Tuning forks received by the 1858–1859 pitch commission · 47
1.3 Table prepared by Jules-Antoine Lissajous and César-Mansuète Despretz presenting the results of the measurement of the forks received · 48
1.4 Prototype of first standard tuning fork · 54
2.1 Alexander J. Ellis's "On the History of Musical Pitch" · 77
2.2 Koenig's forks used at the Ufficio centrale italiano per il corista uniforme · 85
2.3 Illustrated price list of the Hawkes military band instruments · 91
2.4 Musical wind instruments used by David J. Blaikley to study the impact of temperature on pitch · 94
3.1 The great organ of Boston Music Hall (ca. 1870) · 100
3.2 View toward the stage of Boston Music Hall (ca. 1856) · 104
3.3 A flyer for a 20 May 1869 fundraising event organized by the Boston Music Hall Association for a concert entitled "Normal Diapason!" · 107
3.4 The acoustics laboratory at MIT (ca. 1890) · 111
3.5 A table of Levi K. Fuller's measurements from a national survey of instrument builders' forks · 114
3.6 Levi K. Fuller's entire general tuning fork collection, exhibited at the 1893 Chicago World's Fair · 115
3.7a A sketch of the exterior of the Dea-gon-ometer when open and closed · 123

3.7b A sketch of the front interior view of the Dea-gon-ometer · 123
3.8 An "Outline History of Musical Pitch During the Past 200 Years," from John C. Deagan's *R Catalog* · 125
4.1 Werner Lottermoser's bar graph depicting distribution of height of concert pitch · 150
4.2 Reception at Lyne, Capel, Surrey, the home of Captain Evelyn H. T. Broadwood, M. C., M. I. M. T., to delegates of the international committee on the standardisation of concert pitch · 154

INTRODUCTION

Tuning Forks and Global Politics

> Surely, the tone A means one given tone and nothing else.
> And surely no other tone has a right to masquerade as A.
>
> WILLIAM BRAID WHITE

The oboe sounds an *A*. The thin note drifts into the upper reaches of the symphony chamber, reverberating above the heads of the other orchestra members. First the strings join in, gradually sawing toward the same tone; then the woodwinds, brass, and percussion. For several seconds, the sounds that these instruments produce are bending toward the same *A*, but it takes a moment for them to get in unison with one another. Such is the unwritten score that orchestras rehearse at the start of every musical performance, before the concert begins.

These inaugural moments when instruments tune seem suspended outside the historical time of musical styles and repertoires; they form a sort of vestibule, a threshold between the world of natural, unformed sounds, and their display as organized material according to specific rules of melodic and chord progression, in relation to composers' particular aesthetics. Yet there is nothing natural or ahistorical about the tuning of an orchestra. To the contrary, these liminal sounds are already saturated with cultural values and plugged into large-scale techno-scientific networks shaped by global politics. Before she plays her *A*, the oboist, charged with supplying the orchestra's standard pitch, tunes her instrument to an electronic device calibrated to *A* 440 hertz. Now commonly adopted as a point of reference for Western music, this sound only became standard during an international conference held in 1939. Far from being consensual, the adoption of this sonic point of reference was the result of over a century of intense negotiations between nations and across a remarkably diverse array of actors, including musicians, scientists, instrument makers, engineers, and diplomats. Drawing on surviving archival materials and instruments across the world, *Tuning the World* analyzes this historic change and tells the story of how the world's music was tuned. Echoing the inaugural seconds of a symphonic concert, the narrative presented in this book is about the seemingly disorganized sounds of musical instruments initially failing at playing in tune, but nevertheless working toward collective uniformity.

Standards are regulating systems. They are intended to secure uniformity and precision across time and space. From weights and measures that underpin trade to accurate timekeeping that disciplines nations and empires, processes of standardization provoke controversy and require careful negotiation. Standards are often created in reference to the natural world, but regardless of the knowledge used to substantiate their authority, they remain inherently political. Consensus is the most crucial characteristic for securing authority for a particular measurement. Yet of all standards, that of musical pitch has historically been the most subjective and resistant to consensus. Resolving how to quantify something as ethereal as music was immensely difficult. The regulation of what tone should constitute a musical standard raised unprecedented scientific, artistic, social, and political questions throughout the nineteenth and twentieth centuries. At stake were crucial concerns over what music is, over its metaphysical value, its history, and the role that the mathematical and experimental sciences should play in its practice. Should a musical standard be based on mathematical theory, economic practicalities, the aesthetic character of different tones, or the historical connotations of different pitches? Furthermore, the measurement of tones relied on the ear, itself a variable organ difficult to evaluate. In short, musical pitch was not like the meter, or the second, which serves as the base unit of time: as an intangible object, and one saturated with aesthetic values, sound posed additional, highly challenging problems. Through a combination of perspectives from musicology, history of science, and transnational history, this book tells the story of the standardized pitch—of a century-long effort to tune the world, and of the controversies and consequences it entailed. In doing so, it provides new ways to think about how standardization happens, the musical experience, and processes of globalization.

Throughout music's history, pitches were fluctuating concepts: countries, cities, and musical institutions performed music according to their own tones.

The creation of a musical standard in 1939 was the outcome of *longue durée* transformations of the musical field. As long as the voice dominated musical practice, that is, until the end of the medieval period, there was little urgency for a unified pitch. Whereas voices can tune variably within the limit of their respective ranges, the tuning of instruments is much more constrained by their materiality. The idea of a fixed sonic point of reference initially emerged as a response to the development of instrumental music, and specifically to its artisanal and, later, industrial contexts. In the first decades of the seventeenth century, natural philosopher Marin Mersenne and organist and music theorist Michael Praetorius both suggested the adoption of unified standards of pitch—however, these proposals remained purely theoretical and found little response among musical practitioners.[1]

In early modern Europe, tuning was an inherently local practice. Given the centrality of churches in musical cultures, organs were the prime instruments of tuning. Used to accompany sung worship, they were tuned depending on the ranges of the singers' voices. The pitch of these instruments was also subject to financial considerations: since shorter pipes produced higher pitches, organ builders tended to tune their instruments to high pitches in order to save material and thus save money. In addition, because it was much easier for organ builders to tune by shortening rather than lengthening the pipes, organ pitches tended to be raised incrementally each time they were tuned. Over time, the pitch would become so elevated that it strained singers' voices: at that point, the organ pitch would be lowered, and the cycle would begin again.

Nineteenth-century American and European musical cultures continued to exhibit this localized character of pitch. Although musical scenes were increasingly dominated by secular repertoires, they were also dogged by the same sonic diversity that characterized the early modern period. For example, in Paris at the beginning of the nineteenth century, there were at least six different pitches with contrasting standards in use at the three opera houses, the Conservatory, the Royal Chapel, and within the military bands.[2] It was here that the first extensive schemes for regulating musical practices emerged, in the renewed context of industrialization and globalization. With the development of railways and other travel infrastructures, the musical world became more interconnected than ever before, and the lack of a common point of reference for music came to be seen as increasingly detrimental to performers and instrument makers. Addressing this challenge in 1859, France created the first national standard of pitch and fixed an *A* at 870 vibrations or, in modern terms, 435 hertz—a point of reference that several European countries subsequently adopted. (Hertz measures the complete oscillation of a sound wave, from low to high and back again, corresponding to two nineteenth-century French vibrations, which measured half of that oscillation.)

The French adoption of a standard pitch was emblematic of the remarkable variety of actors and interests involved in the definition of such a norm. While other scientific and technical standards were negotiated between mathematicians, physicists, engineers, and natural philosophers, the determination of this artistic standard involved authorities from the musical world. The creation of the *diapason normal,* as the French government called its new standard, therefore goes beyond questions of science, industry, and politics that other standards posed, raising aesthetic considerations as well. This artistic quality made pitch a hard concept to measure and regulate: it was a standard unparalleled for its subjective nature, from both a sensory and an aesthetic point of view. The etymology of the word "diapason," which referred to the vocal range of a singer throughout the seventeenth and the eighteenth centuries, attests to the historical link between pitch and voice. In contrast to other measures, which were designed to disembed measurements from body parts so they could be objectively enforced, the diapason normal had to be implemented in singers' larynges.

During the second half of the nineteenth century, France dominated pitch negotiations, exporting the diapason normal to many parts of Europe, as well as to the United States. In the context of rapid colonial expansion and the resulting circulation of musical instruments and musicians, the French pitch spread to the corners of the earth, shaping global soundscapes. By the beginning of the twentieth century, however, this standard came under increased strain because of the global influence of the United States, which by then had become the world's largest economy. Between World War I and World War II, the United States promoted a different standard, *A* 440, which the leading European nations subsequently selected in 1939. The emergence of this consensus on the eve of World War II not only called attention to the empowerment of the United States on the international stage, but also revealed the reframing of the discussion about pitch around new networks of electroacousticians and sound engineers. In 1955, the 1939 decision was cemented internationally when the International Standards Organization adopted *A* 440 as the global acoustic norm—known thereafter as ISO 16. Its function: "specifying the frequency for the note A" for the "tuning and retuning" of instruments.[3] After a century of intense battles, the regulation of music finally seemed firmly in the hands of Euro-American scientists and engineers.

Despite the official selection of this standard, however, concert pitch remains problematic up to this day. To return to our oboist: if she were part of a twenty-first-century European ensemble, it is likely that the *A* she would produce would be 442, 443, or 444 hertz, which are the Continent's customary frequencies. However, early music performers take more liberty with the official standard. From *A* 392 to *A* 465, they use an ever greater variety of standards as part of their efforts to authentically perform historical works. Moreover, in recent years, this movement has conquered new chronological terrains, as musicians have started experimenting with nineteenth-century standards as well.[4]

Such practices are not limited to the world of classical art music. Since the beginning of the twenty-first century, there has been a proliferation of controversies over *A* 440 in the realm of popular music. Websites promoting conspiracy theories present the international norm as artificial, capitalist, and harmful, and advocate instead the use of *A* 432, often as a means to reconnect with nature or the origins of music. Drawing on some of these claims, in the Netherlands, the centrist liberal political party, Vrijzinnige Partij, argues for *A* 432 as a standard on the grounds that Goebbels was responsible for the adoption of *A* 440, and that this measure causes "disarray" in music and society.[5]

Finally, and most importantly, despite the wide dissemination of European and American musical instruments and repertoires, as well as recordings, audio technologies, and tuning devices, the use of *A* 440 remains mostly confined to Western musical performance, whereas other musical traditions continue to be characterized by various degrees of sonic flexibility.

If musical pitch is still in flux, then, is it really an object worthy of historical inquiry? In other words, given the unsettled character of this measure, does a book about pitch standardization have a *raison d'être*? There are two answers to this question. First, although the degree to which this standard is applied in any society varies greatly, deliberate transgressions against it concern only a limited portion of the Western musical world. In comparison with the diffusion of synthesizers and other musical technologies that silently disseminate the standard across the world, the initiatives of early music performers and countercultures advocating the end of *A* 440 do not undermine the general tendency toward uniformization at play within Western contemporary musical performance. Second, even these reactions against uniformity are a result of the standardization process. In the absence of a dominant order, these critiques would not exist; furthermore, the alternative standards being advocated are a result of the original negotiation process. Like concert pitch *A* 440, they emerged as propositions for a unified pitch during the century of debates examined in this book.

Only by shedding light on the historical processes of standardization can one understand the strategies of diversification that characterize musical performance today. Indeed, the very importance attached to discrete differences in frequency arose only once the actors creating this musical standard realized what was at stake for them. Throughout this book, it becomes clear that beyond its apparent simplicity, the creation of concert pitch offers unique insights into the fabric of musical modernity. Revealing the interconnectedness of science, music, and globalization, its history requires a rearticulation of established disciplinary boundaries and, thus, opens new avenues for research on music and sound.

AN "IMPOSSIBLE" STANDARD?

The creation of a musical standard at once belonged to, and exceeded, broader efforts to introduce uniformity throughout nature and society during the modern era.[6] Questions of science, industry, and politics lie at the center of the history of such standards as the meter, the kilo, or the second. Unavoidably, standardization in music involves both aesthetics and culture as well. If other technical and scientific standards were principally the products of astronomical observatories, industrial factories, and physics laboratories, pitch was inevitably the concern of opera houses and concert halls, churches and radio studios, instrument makers' workshops and music schools. In seizing the initiative for determining the level at which musical pitch should be set, scientists found their credentials disputed by a diverse body of interested parties, including musicians, instrument makers, and politicians. All had different ideas of what constituted a suitable regulating standard. These ideas were shaped not only by considerations about the nature of sound but also by conflicting notions of what sounded aesthetically pleasing, what was physiologically sustainable, and what was historically consistent with the works of celebrated composers.

When we think about tuning, a typical image that comes to mind is that of a piano tuner carrying an instrument, whether a steel tuning fork or an electronic device. However, an enormous number of people were involved in the process that gave rise to this seemingly simple operation. The tuning device used by our tuner was developed through a history of negotiations whose participants included many more players than the mathematicians, physicists, and engineers involved in other stories of standardization processes. These international talks included a broad range of political authorities, from *préfets* and cabinet members to chancellors, kings, queens, and emperors, as well as representatives of different crafts and industries, including makers of diverse musical and scientific instruments, and a broad array of musical parties—composers, performers, instrumentalists, choir conductors, music critics, and directors of musical institutions. In the first decades of the twentieth century, this mix was joined by a new generation of pitch ambassadors: broadcasters, electro-acousticians, heads of standardizing agencies, and members of international organizations. The stakes were high and highly varied for both those attempting and those resisting the standardization of pitch, ranging from financial and industrial concerns to medical and aesthetic considerations, political ambitions, and cultural anxieties. The social heterogeneity of the players explains both the time it took to secure international agreement around *A* 440, as well as the incomplete character of standardization.

The very point of departure for determining a standard pitch was highly controversial, and reflected the variety of social groups and interests involved in the negotiations. For instance, the French physicist Jules-Antoine Lissajous (1822–1880) wanted to set the value of the diapason normal in terms of the metric system, proposing to adopt a *B* producing 1,000 single vibrations, or 500 hertz, as a standard; other actors argued that the voice should be chosen as the ultimate criterion. Still others suggested basing the determination of the standard on the observation of existing tuning practices. This last approach to pitch standardization governed the production of pitch data during different time periods and in various geographic areas, ultimately leading to the creation of a record of diverse sonic worlds, whether past or present. In chapter 2, for example, I consider one of the most impressive collections of historical pitches gathered by the British scholar Alexander J. Ellis, which remains one of the main sources for performance practice studies. Ellis's work reveals how universalist, hegemonic approaches to pitch resulted in an increased awareness of its relativity.

The first impulses for creating a uniform concert pitch reflected the predominance of aesthetic and cultural concerns. These extensive efforts to regulate musical practice were not aimed at unifying frequencies across space to enable the circulation of musicians and musical instruments, but were rather envisioned as a way to prevent changes over time. Although acousticians and musicologists have challenged this view more recently,[7] a fundamental dimension at the start of the negotiations was a shared belief that pitch was rapidly ascending. For instance, in 1858, the composer Hector Berlioz predicted that "pitch—having risen one tone in a hundred years, or half a tone in half a century—would, if its ascending march continued, go through all the semitones of the scale in 600 years, and would necessarily be up by an octave in 2458."[8] As ungrounded and fantastic as this view may seem today, it was a source of deep anxiety throughout the second half of the nineteenth and the first half of the twentieth centuries.

If pitch's alleged ascent was depicted in such dramatic terms, and if, indeed, actors started to look at history to document sonic fluctuations, it is because the past was gaining unprecedented authority within Western music cultures. Since the end of the seventeenth century, "ancient music" had been at the center of the Academy of Ancient Music's activities in London; and after the death of Louis XIV, the Royal Chapel at Versailles kept performing the works associated with the monarch's reign.[9] These trends intensified during the nineteenth century, marked both by a cult of the "Classics"—Haydn, Mozart, and Beethoven—and by a growing interest in music from the Middle Ages and the Renaissance.[10] In the context of this historicization of the musical field, pitch's supposed upward tendency was seen as a threat to the conservation of musical repertoires, especially for opera and other vocal music. For example, audiences wondered how a soprano would be able to sing the high notes of the Queen of the Night's aria in Mozart's *Magic Flute* if pitch continued to rise. In chapter 1, I analyze the creation of the diapason normal in France in relation to the country's most eminent composers' aspirations to return to what they perceived as the golden days of *grand opéra*.

From the mid-nineteenth century, concerns over pitch's historical variations merged with programs of geographic integration. But despite the emergence of new questions, cultural anxieties over the conservation of musical repertoires remained at the core of negotiations surrounding pitch. In 1971, the European Economic Community's Committee of Ministers passed a resolution that epitomized this phenomenon. Entitled "On the Standardization of the Initial Tuning Frequency," this text started with a statement asserting "the need for national and European action to safeguard musical heritage."[11] Throughout the period I consider—an age when music was increasingly seen as an art from the past—pitch evinced a privileged relationship to history. From Bach and Handel to Mozart, Gluck, and Beethoven, old masters played a critical role in the definition of the standard; pitch was envisioned as a regulator of change over time. Today's theories contesting the authority of *A* 440 and suggesting the use of *A* 432 as a return to the origins of sound attest to the resilience of such historically grounded arguments.

From the First French Empire (1804–1815) to the early twenty-first century, the standardization of musical pitch has been a process intended to protect the various components of past musical cultures: compositions, instruments, standards of musical practice, the sonic imagination of great composers. In mid-nineteenth-century concert halls, opera houses, and scientific academies, as well as in post–World War II radio studios, acoustic laboratories, and standardizing agencies, pitch regulations went hand in hand with the invention of musical canons. Ultimately the creation of concert pitch was a result of emerging auditory infrastructures that reconfigured musical spaces—especially the possibility of recording sounds and hearing back-to-back performances from all over the technologized world. However, the production of this standard was consistently envisioned as a way of securing the conservation of musical works from the past.

Throughout the negotiations over a standard pitch, conservationist anxiety intersected with the rise of a new interest in timbre, itself the result of the development of new instrumental genres. This "orchestral revolution"[12] brought with it an emancipation of the various instruments' tone color. When an oboe and a clarinet play the same pitch, they produce a different sound: they have a different *timbre*, or color. At stake in the negotiations was an idea that still shapes musical practice today: that when an oboe plays different pitches, it also produces different colors. More precisely, there was a strong belief that the higher pitches of an instrument added "brilliance" to its sound. Thus, by increasing the tension of pianos', violins', and cellos' strings, or shortening the pipes of wind instruments and organs, one could improve their sonority. One still finds a vivid trace of this notion nowadays in orchestras' tendency to tune just a few hertz above concert pitch (*A* 442, 443, or 444). In the nineteenth century, various categories of musicians already used these strategies. For example, the acclaimed solo violinist virtuoso Niccolò Paganini famously tuned his violin up to a full tone above the standards in use in the orchestras with which he performed. Debates about pitch, in other words, were inseparable from a new attention to tone color and, in particular, a deliberate cultivation of high frequencies for their alleged "brilliance."

For those who feared that pitch's escalation would ruin Western musical heritage, this cultivation of elevated pitch was dangerous, or even criminal. A litany of accusations against various categories of musicians and instrument makers accompanied the negotiations. For example, in 1855, Lissajous blamed brass instrument makers for producing increasingly high-pitched instruments; and a few years later, in 1858, Berlioz similarly accused woodwind builders of "clandestinely raising the pitch" to "give more shine to flutes, oboes, and clarinets."[13] Other actors in the debates incriminated further groups, including piano makers, organ builders, violinists, composers, and singers. In contrast to these conservative comments, many considered that lowering the pitch by implementing reforms was detrimental to musical practice. For example, in 1824, the French music critic Castil-Blaze ironized about the French government's decision to lower the pitch at the Opera, claiming that this would result in violinists playing on a loose string.[14] Similarly, British brass instrument makers lobbied against the adoption of the French pitch until the late 1920s, arguing that it would take away the "brilliancy" of their instruments' sound. Throughout the negotiations, the question of pitch thus crystallized contrasting aesthetic positions, themselves enmeshed with broader cultural prejudices. At the heart of these tensions were conflicting views on the respective role that past and present musical genres and repertoires should play in contemporary musical performance. As some conservative voices summed it up after the Second World War: deciding on a pitch for musical practice ultimately amounted to choosing between Bach and jazz.[15]

In addition to being primarily defined in relation to competing visions of history, pitch standardization was contingent on geographic contexts, a complicating factor that created a lot of misunderstandings. To begin with, the expression "musical pitch" lacks a clear equivalent in French, German, and Italian. In all these languages, the notion has at least two possible translations. While the words "diapason," "Kammerton," and "corista" referred respectively to the standard in use at a given time or place, the words "ton," "Tonhöhe," and "tono" designated the perception of a lower or higher pitch. What is more, in the nineteenth century, the French word "diapason" not only referred to abstract standards in use for musical practice and instrument building—the "convention by which one attributes the name of a certain note to a certain sound"[16]—but also to the instruments that embodied such standards (increasingly, but not exclusively, steel tuning forks), as well as the ambitus of a given voice or instrument (the range of sounds extending from the lowest to the highest note that can be produced).

Ways to quantify and represent pitch varied far beyond linguistic fluctuations. Before the general acceptance of electro-acoustical procedures for sound measurement in the interwar period and the use of the hertz as a uniform unit from 1960,[17] sound measurements were embedded in diverse cultural contexts. For example, in France the use of single vibrations to indicate frequency prevailed, whereas double vibrations predominated in Germany, Britain, and the United States. Tuning procedures were also subject to temperature variation, measurements of which had to be converted between degrees Celsius and Fahrenheit. Determining pitch was also inseparable from counting time, and the second itself was not a universally agreed-upon measure.

A minor error in translation happened to aggravate this metric chaos. In 1860, Britain's Royal Society of Arts published a translation of the French decree on the diapason normal. The 1859 resolution fixed the level of the standard in relation to what was then thought to be Paris's average temperature: 15° Celsius. Tying the pitch to a temperature showed that the decree's authors were aware that temperature, in changing the density of the metal of the fork, had an impact on the way it vibrated and, thus, on its pitch. The translator for the Royal Society misinterpreted the indication of temperature in the text, and for several decades British and American audiences wrongly considered the 15° Celsius accompanying the number of single vibrations fixed for the diapason normal to be the desired temperature level of the room where music was performed. (It actually referred to that of the workshop where tuning forks were manufactured.)

Adding to this metric and linguistic chaos, nineteenth-century pitch negotiations revolved around two different notes: *A* and *C*. In France, Germany, and Italy, conversations started to gravitate around the note *A*, for practical reasons rooted in the reality of musical instrument building—*A* being the note of the open string on a violin. In contrast, however, Britain and the United States followed the long tradition of natural philosophy and music theory, rooted in the practices of organ builders, that used *C* as the foundational note. This represented a major epistemological obstacle, given that the relationship between various musical notes was not fixed, neither historically nor at the time. Equal temperament, the tuning system resulting from the division of the octave into twelve equal parts, only became dominant in the interwar period—and even then, it was only partly realized, as it remains today. Throughout the nineteenth century, the tempered scale coexisted with other tuning systems, including just intonation and various meantone temperaments.[18]

As a result of this variety in tuning systems, the same *A* could produce a multiplicity of frequency values for each note of the scale. Conversely, there were several possible *A*s for any other note of a scale. This is, for example, what the Belgian acoustician Charles Meerens explained, in 1873, when introducing *A* 432 (or 864 simple vibrations) for the first time in the debates. He derived this pitch from *C* 512, a pitch praised for its mathematical quality at the time (it was an octave higher than the mathematician Joseph Sauveur's suggested "ton fixe" *C* 256, a figure adopted by generations of acousticians after him[19]). Meerens detailed how "the acoustic science offered us three numerical values for . . . *A*," including one in Pythagorean tuning, one in equal temperament, and one resulting from the application of a 5⁄3 ratio—by which *C* and *A* were respectively assigned the role of dominant (fifth degree) and third degree in the key of *F* major.[20] Meerens's selection of Pythagorean tuning revealed the contested character of equal temperament at the time, while his justification for this choice—the system's alleged "natural" character, itself a guarantee of its legitimacy—revealed its epistemological, cultural, and sociopolitical implications. Cultural trends such as historicism and globalization in fact embedded concert pitch in ever more diverse systems of tonic organization, thus further contributing to the unsettled character of the standard.

Alongside musical standards, pitch measurements exhibited a local character at the beginning of the nineteenth century. Pitch values were inseparable from the various material and cultural contingencies of their production: to secure the validity of their measurements, acousticians had to describe at length the method and apparatus used to produce them. It is only over the course of pitch negotiations that scholars started to detach pitch data from the specific contingencies of their production, as part of their project to tune the world. Aiming to document tuning practices on a broader chronological and spatial scale, European and American scholars produced assemblages of measurements recorded by different acousticians and, through a variety of practices that were themselves entangled with multiple technologies, worked to redefine pitch as a unified phenomenon. For example, in 1854, the French physicist Charles Delezenne, a member of the Société des sciences, de l'agriculture et des arts (Society of Sciences, Agriculture, and the Arts) in Lille, northern France, produced a table that emphasized pitch's tendency to rise over time, by combining his own measurements of the instruments of Lille's orchestra with that of several scholars from the eighteenth and first half of the nineteenth centuries. Similarly, the French commission justified their intervention by presenting a table demonstrating the escalation of pitch through a list of measurements from various acousticians organized chronologically.[21]

Pitch's unification on a theoretical level seemed to be the condition for its uniformity in practice. But because this unification overlooked the importance of materiality in the production of pitch, this new way of defining pitch made it virtually impossible to implement a standard. The history of pitch standardization reveals the contradictions between universalist conceptions of pitch and the local conditions of its production. Debates about temperature especially crystallized this problem. As mentioned above, in 1859, the decree that imposed the diapason normal in France defined the standard in relation to the sound of a tuning fork at a temperature of 15°C, reflecting both contemporaries' awareness of the impact of temperature on sound and reformers' high standards of acoustic precision.[22] This sealed a complicated relationship between the standard and its various applications that haunted the subsequent decades of negotiations. While assessing the connection between tuning forks and temperature was a relatively straightforward process, things were much more complex when it came to musical instruments in the context of actual performances. As approaches to pitch standardization started to encompass not only the practices of natural philosophers and scientific instrument makers in their laboratories and workshops, but also those of musicians across various settings, as they did from the mid-1880s, the entanglement of musical instruments' varied materiality and design with music's variegated environments made it clear just how daunting the project of unifying sound frequencies would be. This was especially challenging beyond the limits of Europe and the United States' temperate climates, namely in colonial settings where Westerners used sound as a tool to impose their authority over indigenous societies.

Standardizers' awareness of the difficulty of controlling frequencies became all the more apparent with the development of new techniques of sound measurement based on new sonic media. In the period between the two world wars, broadcasting afforded acousticians new means to approach pitch in time, which made them both increasingly aware of its variations over the course of musical performance and eager to control these variations. The more they learned, the more of a problem these variations became—encompassing not only the production of a proper tuning device and the initial tuning frequency of a single instrument or a group of instruments, but also the totality of elements involved in the production of sound, from the temperature of rooms to musicians' routines and the different materials employed to make instruments. Despite increasing sonic control, these techniques have not yet solved the problem that interwar studies started to objectify: it remains impossible to maintain a constant pitch throughout any given musical performance.

In addition to these epistemological and technical problems, I uncover throughout the book many instances of resistance triggered by the making of this norm and the many alterations that followed—from British efforts to liberalize concert pitch, to the American transformation of the standard in relation to the development of popular music and new sound media, to Prussian-dominated German lands' refusal to align with Vienna's tuning practices, to the Australian rebellion against the British Empire. In all of these cases, political considerations intersected with economic concerns: Why invest money to conform to a norm imposed by an exogenous power? Who has the right to direct and govern musicians' techniques and musical institutions' finances?

The history I present shows that the attempt to introduce a unified standard for music has been both laborious and conflicted. One should not, however, always take at face value the arguments that actors mobilized during the negotiations. Among these, for example, were a number of myths. As Berlioz's emphatic statement about pitch's escalation reveals, some fears were not justified—but nonetheless played a driving role. In addition, some parties employed arguments in bad faith. Instrument makers in particular often used artistic or scientific arguments to make the case for universal pitch when their real reasons were industrial or commercial. This attempt to exploit the centrality of cultural concerns inherited from pitch debates during the First French Empire should not overshadow the importance of more recent, shifting economic interests that played a major role in the creation of concert pitch since the mid-nineteenth century: the industrialization of instrument making, the rapid development of global trade and, later, the rise of the broadcasting and recording industry.

Even when standardizers managed to create a consensus around a pitch and embody it through a set of reliable technologies (whether tuning forks, radio signals, electronic tuners, or musical instruments), it was often difficult to make the public aware of, let alone eager to adopt it. Theoretical definitions and physical incarnations of the standard were not sufficient. *Tuning the World* shows that the diffusion of standards involves a complex combination of legal, bureaucratic, social, and aesthetic efforts. Given the different jurisdictions governing diverse aspects of the process, pitch was at once highly mutable and stubbornly immobile.

After World War II, new scientific approaches to pitch started to emphasize the subjective character of the standard. Drawing on psychology, acousticians increasingly differentiated between the phenomenon of frequency and its perception, and began to insist on the impossibility of reducing concert pitch to its techno-scientific specifications. By the time the standardization process was complete, the standard proved applicable in only a fraction of the situations for which it had initially been thought useful. The aesthetic, subjective aspects of pitch have ultimately exercised an equal, if not opposing power to the gradual triumph of standardization.

What might a successful standardization effort look like, given the complexity of the endeavor and the variety of historical actors? Would it consist of the broad distribution of accurate forks, the cooperation of musical parties performing in tune with concert pitch, of audience awareness of the standard? Crucially, to what extent were standardizers themselves able to perceive the transformations of musical soundscapes? In all likelihood, actors' perceptions have been shaped over time by a rising awareness of pitch's heterogeneity, as well as a new sensitivity to the bandwidth defined by particular frequencies. Take, for example, the phenomenon of listeners with perfect pitch, who struggle to hear a "Baroque" performance presented, in its "original" key, at 415—the pitch long deemed appropriate to this era, though it is a full semitone below today's concert pitch. To them, Bach's B-minor Mass sounds as if it was in A-sharp. Such an experience, itself the product of the standardization process, gives a sense of the transformations that two centuries of negotiations have produced on Western ears. Paradoxically, the more standardized musical soundscapes became, the more diverse pitch seemed to be.

How exactly pitch standardization reshaped the way audiences perceive sounds remains somewhat of an open question. The material transformations I analyze often have very little to do with perception—as a BBC engineer noted humorously when the BBC started the daily broadcast of a 440 hertz signal with an accuracy of 1 part in 10 million, this degree of precision lay far beyond what the ear of Sir Adrian Boult, the BBC orchestra's conductor, could hear.[23] While the music's material culture has reshaped the way we experience music and sound, it does not offer a specific picture of the transformations in the history of perception.

SCIENCE, TECHNOLOGY, AND THE MAKING OF MUSICAL MODERNITY

Over the last two hundred years, pitch standardization has shaped all dimensions of musical practice. Ear formation and a good intonation have become the cornerstones of musical curricula and education programs, while the search for stable frequencies has governed the design and modes of production of musical instruments, the acoustics and ambient conditions of music rooms, as well as the techniques and routines of musicians. This transformation has renewed the way audiences experience music across a wide array of musical genres and traditions in several parts of the world. In light of the centrality of pitch standardization in the history of contemporary musical performance, it is striking to note how little attention it has received from musicologists. To some extent, this lacuna results from the standardization process itself. Standards have a tendency to sink "below the level of social visibility."[24] What is more, invisibility is the condition for their success: in the words of historian of science Simon Schaffer, "As for a crime, the secret is not in committing it, it's in making it disappear."[25]

The absence of a thorough study of musical standardization, however, is also the result of a reluctance among music scholars to embrace new approaches from other disciplines within the humanities and social sciences. Over the last thirty years, an abundant literature has taken standards out of their invisibility: scholars in the field of history of science have analyzed the historical contingencies of the creation of these objects,[26] while science and technology studies (STS) have drawn attention to the ubiquity of standards and the ways in which they shape all dimensions of social life, from metrology—the scientific study of measurement—to engineering, medicine, and education.[27] The efforts of these authors have put into question the invisibility of standards and shown rather just how political and cultural they are. Researchers from other disciplines—historians of art and architecture, scholars of media and sound—have followed this lead.[28] In his *Harmonious Triads*, historian of science Myles Jackson has demonstrated the relevance of such lines of inquiry for the study of music. Here, he analyzed the production of standards in light of nineteenth-century Germany's scientific and engineering contexts. The musicological study of pitch, however, has largely remained an object of philological inquiries that aim to restore the standards in use at a given time and place and thus allow the historically "authentic" interpretation of repertoires from the past.[29]

To understand what was at stake in the determination of a unifying musical measurement for pitch, one has to go beyond "purely musical" concerns and consider a wide array of sonic practices. The fixing of an audible point of reference first brought together diverse formal and practical approaches to sound, uniting the intellectual frameworks of natural philosophers and makers of musical and scientific instruments with those of musical practitioners. Moreover, research on musical pitch was rarely divorced from wider scientific interests. Rather, it was typically connected with broader, more crucial, acoustic problems. Lissajous, the brilliant and ambitious young experimentalist who later became famous for his observation of acoustic curves, encountered the challenge of pitch standardization while developing new methods to scrutinize the movements of vibrating bodies, such as strings, plates, or tuning forks. Eager to take advantage of the interest in acoustic demonstrations among French elites to make a name for himself in Paris, he envisaged his work on tuning forks as one of the many applications of a new theory he had developed—indeed, his intervention in pitch debates proved to be an efficient way of advertising his work on vibrations. This case shows that the regulation of pitch was inseparable from the development of laboratory and industrial techniques, as well as from the appeal of acoustics within mid-nineteenth-century European societies. Similarly, Charles R. Cross (1848–1921), one of the first directors of MIT's physics laboratory, played an important role in the standardization of pitch in the United States. Just as with Lissajous, his commitment to musical pitch resulted from his engagement with broader acoustic issues at the time, including debates over the nature of speech and vowel sounds. Cross also benefited from the institutional, financial, and technological support that this research elicited in the United States in the last decades of the nineteenth century.

If research on musical pitch reaped benefits from wider acoustic programs, standard pitches and tuning forks also had an impact far beyond the walls of instrument builders' workshops, concert halls, and other music venues: they were used as disciplining tools in contexts ranging from physiology to psychology, and from neurology to engineering.[30] The tuning fork, for example, was psychiatrist Jean-Martin Charcot's premier instrument to treat catalepsy and hysteria, helped diagnose the causes of hearing loss through standardized tests, and shaped everyday experiences of technology, such as communication through the telephone.[31]

Today, interdisciplinary studies located at the intersection of musicology and history of science are particularly dynamic.[32] But the lack of interest paid by music scholars to pitch standardization as a cultural and political process shows that the field remains primarily dedicated to the study of musical repertoires, styles, and institutions. Such a frame obfuscates some of music history's central features. The Austrian critic Eduard Hanslick, one of the most influential figures of nineteenth-century European music culture, is best remembered for his contribution to the concept of "absolute" music and the myth of music's autonomy. Although much has been written about his aesthetic commitment to the Austro-German musical tradition embodied by Brahms, and his crusade against Wagner, Liszt, and other eminent composers of his time, far less known is his role as one of the most vocal advocates of pitch standardization. Indeed, Hanslick believed such regulations to be the musical equivalent of a well-ordered railway system.

Hanslick's interest in pitch standardization not only reveals the centrality of debates over this question in mid-nineteenth-century Western musical circles, but exemplifies their inseparability from core aesthetic concepts. The critic's commitment to standardization was the product of his concerns over the transformation of European music cultures; conversely, his universalistic views of music owed much to the intensification of musical exchanges across the Continent. Science and technology played a key part in the making of contemporary music cultures. Musical ideas and practices, as ethereal as they seem, are shaped by industrial practices, international trade, and the development of transportation infrastructures. And whereas research where the histories of music and science intersect may seem somewhat irrelevant to the broader field of musicology, it in fact greatly enhances our understanding of past musical cultures. Concert pitch, being an interface between mundane musical practices and larger scientific and technological systems, is a perfect example of scientific insights' relevance for general musicological inquiries.

Given the absence of clear boundaries between music and science and the inextricability of both domains from instrument building in the nineteenth century, the only way to understand the history of musical pitch is to combine perspectives from musicology and cultural history with those drawn from the research undertaken on standards and standardization processes in STS.[33] *Tuning the World* shows how such perspectives can transform our understanding of music. Collectively, this scholarship has demonstrated how standards were inseparable from historical processes that are often seen as part of the "modernization" of societies, such as industrialization, urban growth, globalization, and the development of experimental cultures in the sciences. As the products of highly charged interactions between political, scientific, and industrial networks, these standards connected diverse sites of activities and have played a key part in regulating national and international markets since the beginning of the "industrial revolution."[34]

Just like other scientific and technical standards, the creation of a sonic point of reference figured prominently within the profound transformations of European and American societies in the last decades of the nineteenth century. The Treaty of Versailles in 1919 epitomized this relationship. In article 282, section 22, amid the reaffirmation of economic regulations between Germany and the Allies, which included the unification of the metric system and of pharmaceutical formulas for drugs, is a clause specifying a new musical unit: "concert pitch." This document demonstrates that musical standardization was inseparable from the development of international commercial exchanges well into the twentieth century.

Emphasizing the value of musical standardization in light of broader scientific phenomena does not mean underplaying fundamental differences between musical and scientific approaches to sound. Different actors conceptualize and experience pitch in contrasting ways. What is more, the efforts to standardize pitch involved far more actors and institutions than those involved in previous books that have analyzed standardization processes. This narrative changes the way we think about those projects, complicating previous understandings of the relations between technology, measurement, statecraft, political economy, and culture. Thus, *Tuning the World* shows not only the benefits of tracing the history of pitch standardization within the context of science and technology, but also the utility of music and sound studies for broader historical inquiries. By balancing science's authority over society, music complicates and enriches previous understandings of modernity.

This narrative also changes the way we think about the musical experience. Behind every musical performance lives an enormous apparatus of musicians, composers, instrument makers, institutional directors, acousticians, committee members, bureaucrats, broadcasters, and audiences. These actors work together and against each other to create and experience musical sound. The efforts to invent, resist, measure, and implement a standardized pitch were fraught and only fleetingly successful because the experience of sound is ephemeral and subjective. As much as this story is about the implementation and infrastructures of measurement, it is also about aesthetics.

To be sure, pitch standardization was not an isolated process. Rather, it participated in a broader increase in the attention paid to precision within Western musical cultures. By the time Europe's musical circles started discussing the question of pitch unification, they were also debating the standardization of tempo,[35] of the musical scale, and of musical notes' designation. What is more, it was often the same institutions and actors who advocated pitch standardization and other musical standards. For example, in 1812, when members of the Paris Conservatory began to discuss the question of pitch standardization, they also examined a recent invention intended to regulate time during the institution's performances. Likewise, William Braid White, one of the artisans of pitch standardization in the United States just after the First World War, was also responsible for creating a method to implement equal temperament on pianos.

While recognizing these connections, this book does not provide a full study of increased precision within the musical realm: not only would this largely exceed the limits of a single monograph but, more importantly, pitch distinguishes itself from tempo, the tempered scale, or the names of notes. Considering pitch in light of these other standards actually helps identify its specificity. Tempo, the tempered scale, and notes' designation were unlike pitch in that they did not solicit as much public attention, probably because they did not contain the same density of material and symbolic implications as pitch. Whether or not musicians played in time did not have any consequence for the construction of instruments and was thus of little importance to industrial actors; and while the implementation of equal temperament was of greater concern to these groups, it did not possess concert pitch's political significance—that of being both the symbol of universal harmony and a tool by which one could set the tone of international relations. It therefore does not come as a surprise that ISO 16, fixing concert pitch *A* 440, is the International Organization for Standardization's unique musical standard. This distinction in itself reveals the peculiar position that the standard occupies within Europe's broader culture of musical precision. Pitch, inasmuch as it bridges international politics and large-scale political economies with the most mundane experiences of ordinary musicians and listeners, is an object of unique historical richness.

This book offers a new way of thinking about the materiality of music. In recent years, music scholars have increasingly turned to technology as part of their attempt to break with ethereal conceptions of sound, offering a new organology, studies on embodiment, and scientific and technological perspectives on sound and music-making.[36] In rejecting the concept of music's autonomy, the "material turn" taken by recent musicology has broken with the discipline's long idealist tradition. Yet this trend has only seldom involved an engagement with broader sociopolitical and economic considerations, and may have sometimes paradoxically contributed to reinforcing music's autonomy.[37] Expanding on this research, my approach shows how the study of relationships among music, science, and technology can ultimately reshape our understanding of global history.

PITCH AND "THE WORLD"

The history of concert pitch not only connects musical and scientific questions, but establishes unexpected connections between sound and global politics. Just as for the kilo, the meter, and the second, international and local contexts shaped the manufacturing and dissemination of standards of pitch. As regulating systems, standards are intended to secure uniformity and precision over time and space. Yet different cultures have different understandings of what is precise; and uniform measures have frequently been objects of national and institutional competition, with different laboratories and countries promoting rival measurement regimes.[38] Processes of standardization provoke controversy and require careful negotiation. Unifying measures remain inherently political, whether or not they are created in reference to nature and regardless of the knowledge used to substantiate their authority. Within the natural sciences, the invoking of a natural unit or constant as the basis for a unifying measure has provided powerful strategies for securing a standard authority with scientific and nonscientific audiences alike. For instance, the meter in the late eighteenth century claimed to be a portion of the globe, while the second was based on a fraction of the earth's rotation cycle. Yet ironically, standards often produce variety rather than secure unity. For example, the creation of the metric system during the French Revolution resulted in the juxtaposition of old and new units during the first decades of the nineteenth century.[39] Similarly, the global dissemination of standards of time that began in the middle of the nineteenth century and remained a highly contested process until the aftermath of the Second World War created a mosaic of systems that not only revealed, but enhanced differences in cultural understandings of this concept: we continue to live in a world marked by the coexistence of Gregorian, Islamic, and Chinese calendars, as well as by sustained debates about daylight saving time.[40] In a similar fashion, musical pitch triggered countless processes of diversification and resistance.

Efforts to tune the world from the mid-nineteenth to the mid-twentieth centuries went hand in hand with the constitution of a transnational musical space. As early as 1637, when Mersenne imagined such a standard in his *Harmonie universelle*, it was in reference to a global musical world spanning "Paris to Constantinople," Persia, and China.[41] Similarly, the making of a unified pitch in the nineteenth century was part of the globalization of the musical field as it reacted to the combined impact of industrialization, the internationalization of trade, and colonialism. Standardizers hoped their plans would transcend the limits of their own world: their reforms encompassed the entire globe. Pitch reforms were "universal." Yet a close examination of their efforts reveals that most of the time, "the world" stood for Europe, the United States, and their colonies. What is more, the space that standardizers wished to unify within these limits was mostly that of Western art music, further drawing a geography of elite musical practice. Far from creating a single, unified world, attempts to standardize music produced a multiplicity of distinct "worlds" whose contours differed from one context to the next, and varied over time according to the state of international relations, the signing of trade treaties, colonial and military developments, as well as varied and shifting cultural values.

Although *Tuning the World* mostly focuses on Europe and the United States and not the world, it shows that the standardizing of sound undertaken by these nations was *world-making*. The ephemerality of pitch, whatever its material forms, motivated not just standardization but expansion. The impulse to standardize pitch became a type of proxy for the imperial ambitions of Western nations. Relating its story provides a history of how they defined their worlds.

The world of the standardizers changed throughout the negotiations, from a single musical site in the first years of the nineteenth century—the Paris Opera—to entire nations in the second half of the nineteenth century, to internationalist projects from the 1880s onward. At first limited to Europe and its colonies, it came to span the Global South after World War II. Efforts to tune the world and the power relations that resulted therefrom took on many forms, from the French government's imposition of a national measure on its own *départements*, to Austrian and British efforts to unify frequencies across their vast empires, and from Europe's exertion of influence over the United States, to American imperialism and Western modes of global governance. Tracing pitch standardization requires us to combine methodologies—including global, postcolonial, and transnational histories—and to vary constantly our levels of analysis. In working with actors' stories and conceptions of their own "worlds," this book makes an intervention in ongoing dialogue about what constitutes the "global." Rather than mobilizing a preconceived idea of "the world," I show how music and sound can help us recover the historical, shifting, and competing notions that surround the concept. In doing so, I demonstrate how universalist ideals played a key role in shaping the emergence of Western nations.

Standardizers' universalist ideas of sound and music transformed the world's soundscapes, and this book contributes to the conversations that composer Murray Schafer inaugurated with his influential *The Soundscape: Our Sonic Environment and the Tuning of the World.* In this work, Schafer used the concept of "soundscape" to draw attention to the sonic implications of modernization. Claiming that our sonic environments were undergoing a rapid transformation, Schafer warned of the detrimental effect that noise pollution imposed on industrialized societies and called for a better management of the world's sounds as well as for the development of acoustic design. To do so, Schafer promoted the emergence of environmental acoustics as a way to determine which sounds were worth preserving and which to eliminate. Referring to the recent expansion of music to include all sorts of sounds (such as the works of John Cage), and contemplating the world as "a macroscopic musical composition," Schafer considered it the musician's responsibility to provide a new orchestration of our environments.[42]

More than forty years since the publication of Schafer's program, my book provides a prehistory of the state of affairs that Schafer analyzes in his. I show that modernization had an impact on the world's sound environments long before the emergence of noise as a sociopolitical category. I also provide a new definition of "the soundscape." Rather than referring to the sounds already extant in our modern spaces, I focus on sound fabrication. Throughout this book, I define the "soundscape" as a hulking infrastructure of sound creation and maintenance. In doing so, I highlight the political implications of the notion, by showing the co-construction of soundscapes, world-making, and imperial conquest.

Beyond building on Schafer's seminal work, my work further contributes to ongoing reconfigurations in the field of sound studies. While previous books on the relations between music, science, and technology have mostly focused on a national scale and a few select countries, including Germany, the United States, and Britain,[43] the history of pitch standardization reconstructs the significance of and manifold connections between different territories. This redefinition of the frame of study does not merely add new contexts to the map of studies of sound and science but, more importantly, contributes to an ongoing redefinition of their scope of inquiry. Scholars' focus on the United States and Germany has indeed contributed to a homogenization of the questions within these fields, and a related tendency to universalize problems that were instead specific to these two geographic areas. Drawing on this observation, several scholars have called for an expansion of sound studies' field of investigation.[44] *Tuning the World* contributes to this tendency by uncovering the variety of Western sonic epistemologies. Designed to discipline and unify sonic and musical practices, whether as a historical phenomenon or an epistemic tool, concert pitch is thus a fecund site to start recovering the great plurality of sound, as both a material and an immaterial object. Far from simply describing a process of sonic uniformization, I revive the forgotten voices and propositions of actors involved in the making of this norm, and thus ultimately provide grounds for a "de-tuning of the world."

While the limits of the worlds that this book examines were in flux, the negotiations around *A* 440 nevertheless resulted in the creation of global lines of demarcation that continue to shape research on music up to this day. In the context of a growing awareness of the relativity of Western musical sounds and evidence of how musical cultures are historically and geographically contingent, standard pitch, like the tempered scale and tonality, became reified as a natural feature of European music. The production of knowledge borne along the negotiations to document pitch's historical and geographic fluctuations has informed the way Western music scholars and musicians think about, and practice, music. The work of the British scholar Alexander Ellis at the end of the nineteenth century, examined in chapter 2, shows how pitch standardization durably shaped scientific approaches to music. A scholar of mathematics and philology, Ellis conducted an extensive study of musical pitch on a global scale that inspired both the first generation of comparative musicologists around 1900 and the pioneers of the "historically informed performance" movement after World War II. He thereby ultimately laid the groundwork for two of musicology's branches, located on each side of the discipline's main divide between "Western" and "non-Western" musics: performance practice studies (itself a part of historical musicology) and ethnomusicology.[45]

In unraveling the variety of Western epistemologies, this book also avoids overly simplistic narratives based on binary oppositions between Western and non-Western music and sounds. Crucially, by revealing the chaotic state of Western musical infrastructures far into the twentieth century, it sheds new light on the notion of Western music. To a large degree, "tuning the world" remained a mere wish. In using this phrase, I thus do not mean to reiterate the hegemonic and unrealistic gesture of standardizers. Rather, I aim to draw attention to the tension between the vision of those eager to introduce a global norm and the diversity introduced by the myriad responses to their project. In showing Europe's and the United States' repeated failures at securing global sonic uniformity, I ultimately challenge some of our most common assumptions about Western music. As I show in the book, the expression "Western music" was forged as a performative notion in order to soothe Euro-American anxieties amid a rapidly changing musical world. In using this concept without proper contextualization, music scholars tend to accomplish in their writings what historical actors failed to do in reality, thus ultimately reinforcing the hegemonic gestures they aim to critique.

Further, my taking an empirically grounded approach allows me to revise some of our most common assumptions about global politics. In particular, since pitch is located at the intersection of music, science, and industry, the negotiation of this standard involved actors whose role in the history of international relations has remained invisible so far, among them broadcasters, electro-acousticians, and instrument makers. Turning to these figures, I explain how future enemies were able to agree on a norm synonymous with universal harmony just a few months before the outbreak of World War II. The year 1939 is not an isolated example: throughout this book, it becomes clear that music has often played a role in international exchanges during the modern era. And although it bears the mark of greater geopolitical turmoil, the history of musical diplomacy also reveals how countries were sometimes able to reach agreement amid strained relations. Whether we focus on British attempts at introducing the French pitch in 1860 in the midst of political anxieties over Napoleon III's autocratic regime, German efforts to spread the diapason normal across their territory despite increasing rivalry in the 1860s, or the adoption of a "German standard" in the United States after the two countries went to war in 1917, the history of musical standardization sheds new light on the history of international relations.

The five main chapters of this book track the standardization of musical pitch from Second French Empire to post–World War II Europe and the United States. Going back to our oboist, the book follows her across multiple—aesthetic, scientific, industrial, and political—transformations, showing how her seemingly tiny, mundane gestures are actually the result of a larger history of standardization on a global scale. The oboist does not simply produce a concert pitch with her instrument but first adjusts the pitch of her oboe using a small tuning device. Her effort serves to mediate a specific, materialized sound to the rest of the ensemble. Why is it considered necessary? This is what I examine in chapter 1, by looking at the creation of the French diapason normal in 1859. I unpack the interplay of diverse epistemologies, practices, disciplines, and social fields that shaped this point of reference, and trace its political foundations back to Napoleon III's centralized, autocratic government. Chapter 2 studies the international dissemination of this French standard into the first decades of the twentieth century, asking what happens if our oboist is traveling and must tune to a different concert pitch, one used by local ensembles. The unification of the world's musical soundscapes raised innumerable political, bureaucratic, economic, and material challenges, revealing sound's remarkable tendency to indiscipline. In chapter 3, I turn to the oboist's practice of retuning after intermission or between pieces as the concert hall heats up or as additional instruments join for specific pieces. I show that this gesture is tied to two broad reconfigurations of the field of negotiations: the empowerment of the United States on the international stage and the rise of the entertainment industry, which, together, led to the replacement of the diapason normal with *A* 440. Chapter 4 explains how the oboist came to use an electric tuner instead of a steel tuning fork. It shows that the production of electroacoustic signals to embody the standard is inseparable from the development of a new diplomacy of pitch, enmeshed with the traumatic memory of World War I and Western engineers' desire to keep the world at peace. Finally, chapter 5 revisits our oboist's practice of tuning her own instrument even when performing alone. The last stage of pitch negotiations, amid a dramatic reconfiguration of power relations and sound knowledge, revealed the fundamentally subjective character of the standard. Ultimately, the only soundscape one experiences is one's own.

From a purely sonic point of view, *Tuning the World* investigates the shift from 435 to 440 hertz: an almost negligible difference in terms of hearing, yet one that embodied enormous cultural tensions and historical change. The five vibrations separating the diapason normal from the American standard adopted in 1955 are the audible manifestation of standardizers' struggles to implement a universal point of reference. Rich with a long history, our current concert pitch continues to resonate with the voices and actions of those actors—human and nonhuman—involved in the making of this unique standard.

All in all, this book shows that far from being ethereal, music and sound are shaped by large-scale techno-scientific and industrial systems. At the same time, culture and aesthetics, in fueling many different forms of resistance and appropriation, exert their own fine-tuning power upon the world's political economies.

CHAPTER ONE

Tuning the Nation

Aesthetics, Science, Industry, and the French Pitch

> What! You will submit, to the same rule, the vocal cords and the violin string, mucous membranes and copper?
>
> CHARLES-JACOB MARCHAL DE CALVI

What drove the first large-scale effort to tune the world? Why did a government take action to unify musical soundscapes worldwide? And why did its decisions create long-lasting, intense controversies? In answering these questions, this chapter highlights the historical contingencies that underpinned the creation of the first standard pitch, explains why so many parties got involved in attempting to define it, and details what was at stake for them in the process.

On 16 February 1859, France was the first nation to introduce, by a decree, the use of a national standard of musical pitch for all state-funded and state-controlled institutions. The text of the imperial decree stated that the new standard had to be a quarter of a tone lower than the one in use at the Paris Opera and set the standard tone, an *A* above middle *C*, at 870 vibrations, or 435 hertz. Three weeks later, a professor in the Faculty of Medicine in Paris, Charles-Jacob Marchal, published a "free reprimand" in the columns of the music journal *La France musicale*. Dr. Marchal expressed his disappointment in his government's decision and justified his intervention on medical grounds. On the one hand he disapproved, as a physician, of the choice of what he considered an insufficient remedy to the damages caused by high pitches on the larynges of vocalists. In Marchal's opinion, a pitch of 435 hertz would not protect voices from the "excesses of tonality." On the other hand, as a lover of music, Marchal worried that lowering the pitch would eliminate its expressive power. Sharing the widely held assumption and aesthetic premise that "the higher the pitch, the brighter the sound,"[1] Marchal demanded "emotion whatever its cost, without any regard for the vocal cords of my neighbor."[2] To reconcile his physiological and artistic concerns, Marchal suggested the adoption of a "dichotomous solution": fixing not one, but two different standards. He summarized: "There is a vocal and an instrumental music. Can, and should, the conditions of tonality be the same whether it applies to vocal, instrumental, or vocal-instrumental music?"[3] In other words, should one "submit, to the same rule, the vocal cords and the violin string, mucous membranes and copper?"[4]

Marchal's reprimand reveals both the potential of pitch to provoke antagonistic recommendations from the diverse parties who work with it, and how these differing conceptions were entangled with specific social practices. Marchal may have exaggerated the extent of his internal conflict and used this self-division as a rhetorical device to write an appealing feuilleton and secure the sympathy of readers, but the categories of his musical taxonomy—vocal and instrumental—correspond to two distinct perspectives on pitch: the medical and the aesthetic. They also reflected separate sets of skills and experiences: those of the physician and those of the music lover.

This chapter examines French attempts to standardize pitch. I show that the diapason normal was a hybrid measure, combining scientific and industrial ideals of universality and uniformity on the one hand, and France's dominant aesthetic and musical hierarchies on the other. In comparison with other scientific and technical standards such as the meter, the kilo, or the second, music posed additional problems that were at odds with the very notion of sonic uniformity. Where Marchal's questioning led him to suggest a dichotomy of standard pitches, commentators from other traditions, using diverse bodies of knowledge and techniques from music, physics, physiology, and instrument making, pointed to an even wider multiplicity of standards. Like the physician, these actors involved in the production of France's unified pitch emphasized the importance of securing a norm, but diverged greatly over what it should be—differences that arose from fundamental conceptions of the very nature of pitch, sound, and music. Some, like Marchal, suggested this diversity of approaches be reflected through the adoption of multiple standards. Emperor Napoleon III's government, however, attempted to combine these conflicting views and claims into a single norm. Beyond the negotiation of a specific numerical figure, tuning the nation was therefore a process by which a heterogeneous network of actors negotiated an equally diverse array of theories and practices of pitch.

The French state's interest in sonic standardization resulted from the convergence of two separate traditions for pitch reform, one led by acousticians and the other initiated by musicians. To be sure, the boundaries between these two worlds were porous. Acousticians drew on musical theories and technologies in their experiments with sound, while musical practices were shaped by scientific and technical transformations. In addition, actors of these two worlds often did not have the same views, and the groups "acousticians" and "musicians" include diverse subcategories such as scientific instrument makers, mathematicians, and physicists. Music instrument builders spanned both of these worlds, fostering intense exchanges between the two. Yet despite the cross-fertilization of these fields, pitch was never a unique object. Rather, it referred to a plurality of conceptions and experiences. Standardization efforts that aimed to improve experimental procedures or those meant to enable musical performances relied on different premises and operations, and pursued distinct goals. For acousticians, the production of a normal pitch was largely informed by mathematical or physical knowledge, required the widespread distribution of standard instruments and procedures, and was seen as a means to facilitate communication within international networks of scholars as well as to increase precision in experiments, which relied on tuning forks and accurate sound measurement. The standardization of musical practices, on the other hand, was usually based on aesthetic considerations, although it also involved the modification of musical instruments. The creation of a standard pitch was a way of addressing the challenges faced by performers who traveled across territories and used different musical systems, as well as a response to a perceived escalation of pitch over time that was viewed as a cultural threat. In other words, science and music referred to distinct, if connected, contexts that themselves created differing systems for understanding the production of pitch.

The composition of the 1858–1859 commission that proposed the diapason normal reflected this dual character of pitch: it was composed of representatives from the worlds of science and music. What were their respective roles in the creation of this norm? How did the differing epistemological understandings of pitch interact during this process? A close examination of how the standard came about shows that although the material construction of the diapason normal relied on cutting-edge achievements in the field of sound metrology, its theoretical definition was informed by essentially musical considerations. Indeed, the fact that the agreed-upon standard lacked a stronger connection with arithmetic undermined its pretensions to universality and subsequently proved to be an obstacle to its national and international acceptance. In addition, the technological, economic, and sociopolitical issues triggered by attempts of governments as well as scientific and musical institutions to push for its use engendered resistance on a multiplicity of levels.

French authorities in the mid-1800s imagined themselves at the center of the production of universal norms and standards. In the wake of the successful creation of the metric system by the French state during the Revolution and its gradual adoption internationally over the century following, the French state gained considerable power over science and commerce. The establishment of the diapason normal now promised to secure for the country the role of a cultural leader as well. Yet music, as an art invested with considerable authority over society and organized according to its own principles and hierarchies, would not prove as easy to regulate as other realms of metrology.

SCIENCE, MUSIC, AND THE DOUBLE ORIGIN OF THE FRENCH PITCH

On 2 May 1855, Jules-Antoine Lissajous, a young and ambitious high school teacher who specialized in the study of acoustics, delivered a vibrant lecture on sound, in which he called for the organization of an international congress to stabilize and unify musical pitch throughout the world.[5] A regular, but unlucky, candidate to the Académie des sciences and an active member of Parisian scientific and industrial networks throughout his career, Lissajous showed himself eager to use acoustics' growing popularity among European elite circles to establish his reputation as a physicist.[6] After all, sound's potential to secure public interest had long been demonstrated, and was aroused again by the publication of German experimentalist Ernst Friedrich Chladni's (1756–1827) influential treatise *Die Akustik* (1802). Chladni had embarked on a tour of Europe, performing spectacular demonstrations with vibrating plates that met with great enthusiasm, especially in Paris.[7] In 1845, the scientific instrument maker Joseph Marloye and the natural philosopher César-Mansuète Despretz attracted similarly great attention when they exhibited at the Sorbonne a series of "giant and midget tuning forks" that produced from 20,000 to 73,000 vibrations, creating an aural experience encompassing the broad range of human hearing. As a critic reported, this "offered a curious spectacle, for scholars as well as for the gentry," an effect comparable to "Quasimodo's febrile and exalting musical experience in [Victor Hugo's] *Notre-Dame de Paris,* that of becoming intoxicated with the bursting and stunning of his dear bell."[8] Historian Gabriel Finkelstein has argued that during the nineteenth century the French capital was the "Broadway of scientific performance."[9] Acoustics provided the city with some of its most intense highlights.

Lissajous had a clear sense of acoustic phenomena's power to secure public attention, and to make a case for a universal pitch, he arranged for his appearance to take place before a meeting of the Société d'Encouragement pour l'Industrie Nationale (SEIN), a governmental body Napoleon Bonaparte had established to promote innovation through philosophical inquiry, commerce, and banking.[10] Citing the precedent of the metric system, he emphasized the prestige the empire of Napoleon III could expect from adding yet another standard to its celebrated list of measures. Lissajous boasted how

> France now possesses a complete and authentic collection of various measures. The care brought to the confrontation between the secondary standards with the prototypes stored at the archives, that is, the means employed to continually control the exactitude of commercial and industrial measures, ensure the indefinite conservation of this admirable system. It would be desirable that the same principles be applied to the establishment and the maintenance of pitch which serves, in some sort, as a sonic unit and for which there is no official standard up to today.[11]

With its claim to be a division of the distance between the Earth's pole and its equator, the meter not only offered Lissajous a scientific precedent for a measurement of music, but encouraged him to propose a reform of pitch based on the decimalized metric system.[12] Although he recognized that musicians might well be the best judges of what such a pitch should sound like, the physicist suggested the standard of a *B* above middle *C* at 1,000 vibrations per second, or 500 hertz.[13]

The lecture's uniting of scientific and artistic considerations was typical of the SEIN, where he was making his first appearance as a new member of the society. The body was eager to fashion Paris into a center of scientific and musical technologies, and had particular interest in the economic and technical value of organs and pianos, as well as the acoustic and musical questions involved in their manufacture.[14] In addition to the SEIN's tradition of musical expertise, Lissajous's call echoed the body's broader promotion of metrological unification. Throughout the nineteenth century, the society would be a vocal advocate for international measures.[15]

Although consistent with the SEIN's dual commitment to musical technologies and metrological unification, Lissajous's proposal was surprising in its form, which took the shape of a lecture on French opera. Drawing on measurements he had made himself or borrowed from other acousticians, Lissajous asserted that pitch had risen by approximately one tone since the reign of Louis XIV. This king was associated with the formation of the lyrical genres that still dominated French musical culture and epitomized France's musical grandeur: the *tragédie lyrique* and *opéra comique*. To make his audience fully aware of what this meant, Lissajous offered them an audible display of this historic change. Taking as his examples his forerunners Chladni, Marloye, and Despretz, the physicist appeared in front of his audience armed with seven tuning forks installed on wooden resonators, which he introduced as pieces of aural evidence. Sounding them successively, Lissajous delivered a spectacular lecture on the "ascending march" of pitch since the late seventeenth century.[16] Together, the forks represented "the main steps" of this upward journey: the first one, producing 810 vibrations, corresponded to the "last years of the reign of Louis XIV," while the last, giving off 898 vibrations, stood for the Opera in 1855.

This audible display of pitch's historical escalation was meant to alert his audience to a problem that, Lissajous argued, had nothing less than "disastrous" consequences. His lecture echoed the concerns of musicians and music critics, including Marchal, who had long warned the government about the damage to singers' voices caused by high pitches. As Lissajous put it:

> How many beautiful voices have been broken before going on stage, and how few survive the theater's demands! So many prominent singers spend half their artistic life ruining the means that nature gave them and the other half hiding, through art, the early ruin of an organ that no longer matches their talent. This is the disastrous consequence of pitch ascension.[17]

To complete this alarming picture, Lissajous presented the reasons for such a catastrophic situation. He laid the responsibility for this rise on the builders of brass instruments, organs, and pianos, along with unscrupulous makers of tuning forks. He accused the former of cultivating higher pitches both to improve the sonority of their instruments, and to improve their profit margin by reducing the size of their pipes or strings, thus saving money on metal. Lissajous further decried "the method vulgarly employed for the adjustment of tuning forks . . . with the help of a file," a process that resulted in heating tuning forks, thus decreasing their pitch. To be specific, he detailed that "when [the tuning fork] is adjusted, it is in tune with the primitive tuning fork; but . . . when it cools down, it will rise in pitch."[18] As a representative of the natural sciences who cultivated an interest in acoustic apparatuses, he himself was developing a new, far more precise way of adjusting tuning forks. One of the footnotes that appeared in the subsequent publication of his paper mentioned his ongoing experiments with sound and announced that he would soon present a method enabling the implementation of his proposed reform.[19]

Lissajous's lecture triggered the opening of major pitch negotiations both in France and internationally. It was also prelude to a series of presentations during which Lissajous promoted his new method of rendering vibrating visible. The fixation of a unified pitch was only one of the applications the physicist envisioned for his new method, but one with particular potential.

Between 1855 and 1856, Lissajous delivered lectures across France and abroad in the hope of securing an appointment at the Académie des sciences. Throughout these demonstrations, he presented details of several experiments intended to "highlight the vibratory movement of bodies."[20] This method allowed him "to study, without the help of the ear, any kind of vibratory movement, and as a result, any sort of sound."[21] Lissajous set up a method that worked like a microscope for the observation of subvisible phenomena, magnifying the movements of a body by relying on well-established laws of visual perception. By "stick[ing] a small polished plate on the end of one of the prongs on the convex face of the tuning fork," he created a "mirror" that captured "the reflected image of a candle placed a few meters away." This caused the tuning fork to vibrate, revealing "immediately the image enlarged in the sense of the prongs' length"[22] (see fig. 1.1).

Lissajous's experiment addressed the main epistemological challenge faced by acousticians in their attempt to study vibrations: the imperceptibility of the phenomenon. As Lissajous observed, "the vibratory movements that determine the production of a sound take place with such speed" that one "doesn't have time to catch them."[23] In his *Mémoire*, Lissajous confessed that he had been "struck" by the display mounted by the physicist Paul Desains when he "projected" figures on the wall of the Faculté des Sciences' amphitheater in Paris.[24] Lissajous's experimental procedure was designed to make vibrations visible for "a whole crowded audience," "from a great distance."[25] As such it provided yet another of the type of spectacular public demonstrations of French academic culture that had become popular with

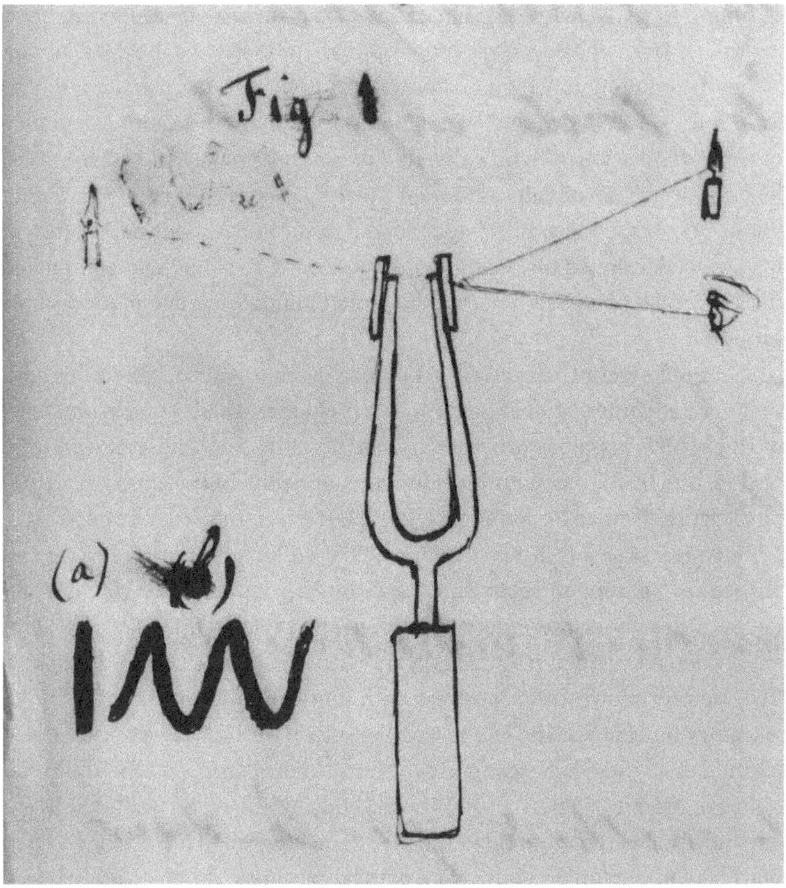

FIGURE 1.1. Jules-Antoine Lissajous, *Mémoire sur l'étude optique des mouvements vibratoires* (1857), 2. Académie des Sciences.

Parisians. For example, the physicist Claude Pouillet (1790–1868), who had supervised Lissajous's dissertation, used a giant electrical magnet capable of carrying 2,500 kilograms. "Every year," the science popularizer Louis Figuier reported, "one sees the magnet support a platform on which seven to eight students come and sit."[26]

Since the seventeenth century, natural philosophers had sought to demonstrate the phenomenon of sound vibrations by making them visible. At the beginning of the seventeenth century, the scholar Marin Mersenne (1588–1648) explained how experimentalists could count the oscillations of a suspended vibrating string by observing its movements while keeping track of time.[27] Thomas Young had examined the strings of a piano, and Charles

Wheatstone focused his attention on vibrating plates covered in various media like sand or mercury. Lissajous insisted on the interchangeability of this object with "any kind of vibrating body,"[28] and decided to utilize the tuning fork after making initial trials with plates.[29] His decision reinforced the new centrality of this tool within Europe's experimental cultures, a factor that would have far-reaching implications for pitch negotiations. Lissajous explained his preference for using what he conceived to be both a scientific and musical object on the grounds that the tuning fork was, "of all the bodies one can make vibrate, the most practical to experiment with."[30] Lissajous's activities built on an acoustic tradition that had turned the tuning fork into a crucial instrument of physical experimentation.

Just as Lissajous was not the first to visualize vibrations, he was following in a long intellectual tradition of natural philosophers who had suggested the creation of a standard pitch. In their attempt to achieve ever-increased precision in their experiments with sound, scholars in this field had long suggested adopting a sonic measure. As early as the beginning of the seventeenth century, Mersenne had envisioned such a standard, which he argued would allow "all the musicians in the world to make the same musical work sing according to the composer's intention."[31] In order to create a science of sound capable of rivaling the already established field of optics, Joseph Sauveur similarly argued that it was desirable to identify a "fixed pitch" independent of the great variety of cultural and technological conditions characteristic of musical instruments. In 1700, he proposed organizing musical intervals around the arithmetic principle of a pitch sounding at 100 vibrations per second—which would have been low for a musical standard, given that it roughly corresponds to the *G* belonging to the second lowest octave on a piano tuned to *A* 440.[32] Sauveur later revised his proposal based on arithmetical proportions: in 1713, he suggested the organization of sounds around a *C* of 256 vibrations, an abstract figure fixed, once again, in reference to arithmetic: it corresponded to the power of 2 that was the closest to the pitch of an organ pipe that Sauveur had measured, which produced 243.2 vibrations.[33]

In 1800, Ernst Friedrich Chladni, the German acoustician who had been the first to demonstrate the acoustical qualities of tuning forks, drew on similar mathematical principles to recommend the adoption of *C* 128, that is, the octave below Sauveur's "son fixe." He assigned this value to the lowest *C* on a keyboard or a cello, and took it as a point of departure to derive successive octaves of *C*, including 256 and 512 (as well as the fundamental, inaudible *C* 1). Likewise, in the 1820s, the German acoustician Wilhelm Weber also became interested in the concept of a *Normalton*. A scientist well acquainted with Chladni and his work, Weber conducted numerous experiments on reed pipes, as part of a project to improve the art of organ making. Despite their shared efforts to create a standard that was equally acceptable to acousticians, instrument makers, and musicians, the calls of natural philosophers and mathematicians remained unheard throughout the musical world.[34]

It was not just mathematicians who had proposed pitch reform. Skilled artisans also engaged with this question, such as the silk manufacturer Johann Heinrich Scheibler, who conceived of a method that allowed for unprecedented precision in tuning.[35] Drawing on his experiments with Viennese pianos, Scheibler promoted a standard pitch of *A* 440, which he asserted to be the average number of vibrations produced by the key *A* above middle *C* of these instruments. In 1834, he presented the results of his measurements to the physics section of the Versammlung deutscher Naturforscher und Aerzte in Stuttgart and suggested choosing *A* 440 as a universal standard. Although Scheibler's norm was adopted by a number of Viennese piano makers, it was not enforced in the German states as a whole, and barely reached other territories. Nevertheless, the so-called "Stuttgart" or "German pitch" was to play a crucial role in France's effort to establish a musical measure, and in 1939 would indeed eventually become the international standard.

The work of Chladni, Weber, and Scheibler was well known among acoustic circles in France, where it informed scientific definitions of a sonic point of reference in the decades preceding Lissajous's lecture.[36] In 1832, the engineer and mathematician Gaspard de Prony suggested the adoption of a pitch derived from Chladni's suggested standard, a *C* producing 512 vibrations, as a point of reference for physicists and musicians.[37] In 1834, Marloye participated in the congress at Stuttgart and advocated the adoption of the same figure as a European standard.[38] Although his proposition did not receive as much attention as Scheibler's, Marloye continued to sell tuning forks of *C* 512 throughout his subsequent career, conceiving of these instruments as devices that might facilitate transnational exchanges between acousticians and increase the precision of their experiments.[39] Marloye's activities were emulated in the Parisian workshop of Secrétan and Lerebours, who subsequently designed the prototypes of the French pitch tuning forks of 1859.[40] In their 1853 *Catalogue*, these skilled artisans advertised a fork similar to Marloye's. They boasted that this instrument, which they described as a "diapason normal," was the most "convenient for calculation."[41] Their catalog demonstrated how, in the world of instrument making, precision and normalization carried epistemological implications, such as the possibility to replicate the same experiments in various places and at different times. These values were playing an increasingly central part in commercial and advertisement strategies.

By the time Lissajous addressed the SEIN, tuning forks were being utilized within Paris physics and physiology laboratories for many nonmusical concerns. Confirming this separation of acoustic experiments from purely musical interests, Lissajous expressed the hope that the method he presented would find applications far beyond the realm of musical acoustics, including especially the study of optical phenomena.[42] In this regard, Lissajous's approach to acoustics was exemplary of the subdiscipline's status in mid-nineteenth-century Paris: in physics textbooks and classes, as well as in scholars' experiments, acoustics typically served as an introduction to more general physical problems in nature, such as heat and light.[43] As with later efforts to regulate pitch throughout the nineteenth and the twentieth centuries, Lissajous's engagement with musical pitch derived from a more general examination of nature, whereas some of the technologies and techniques that scientists employed had been imported from the field of musical practice or instrument building, such as tuning forks, organ pipes, and beat counting.

French efforts to establish a national standard pitch brought to the fore the great differences between the ways in which musicians and acousticians approached these objects and practices. In particular, the involvement of physicists within these theoretical and technical discussions proved transformative. The question, then, was: with all these contrasting interests and values, to what extent could these scientific conceptions and practices of pitch be brought in line with those of musicians?

Musicians as well as acousticians expressed an eagerness to unify pitches throughout the first half of the nineteenth century. Although less exaggerated than in previous centuries, pitches were still inconsistent throughout Europe, even at a local level. In Paris at the beginning of the century, there were at least six different pitches with different standards in use at the three operas, the Conservatory, the Royal Chapel, and within the military bodies. In his 1855 lecture, Lissajous challenged the concept of an "opera pitch" altogether, by emphasizing the fact that this institution, the very pinnacle of France's musical scene, did not even possess its own *étalon* or standard tuning fork.[44]

Pitch discrepancy was manifest when musicians circulated from opera houses and concert halls to churches. Pursuing the traditional distinction between *Tons de chapelle* and *de chambre*, organ makers typically tuned their organs to frequencies far removed from the ones used in other music venues.[45] The composer Hector Berlioz, for instance, complained that the organ of Saint-Eustache in Paris, where he directed his *Te Deum* in 1855, was tuned "a quarter of a tone higher" than instruments from nearby theaters that took part in the performance and that, as a result, "it was impossible, despite lengthening the sonic tubes of all instruments, to have the instrumental mass agree with the new organ, finished just three years earlier."[46] In this case, the even greater acoustic challenge lay in the distance between the orchestra and the organ, located as they were at both extremities of Paris's second largest church (the organ above the Western door, and the orchestra in the nave, in front of the choir), rendering the perception of a unified performance altogether difficult.[47]

If anyone was particularly well placed to discuss the variations of pitch between the different spaces of French music, it was Berlioz. Emulating the model of German and English festivals, the composer introduced the practice of "monumental"[48] performances in Paris in the mid-nineteenth century. However, he also challenged the concept of pitch discrepancy; or rather, rejected the idea that this represented a major practical concern. In a feuilleton he wrote for France's leading daily newspaper, the *Journal des débats*, he explained to the public that

> if there was in Paris, as it is so often said, a great difference between the pitches of the Opera, the Opéra-Comique, the Théâtre-Italien, and military bands, how would the orchestras of seven to eight hundred musicians—which I have so often directed, in the large halls of the Champs Élysées, after the Exhibitions of 1844 and 1855 and in the church of Saint-Eustache—have been possible, since the elements of these musical congresses were necessarily formed by nearly all the instrumentalists from the numerous musical bodies of Paris?[49]

Referring to the festivals he organized within the context of industrial exhibitions, the composer argued that the variations between pitches employed by different orchestras only represented "nuances" and did not prevent a musician from "bringing sometimes these ensembles together, by means of certain precautions, in a great instrumental mass that is satisfyingly tuned."[50]

For musicians, the real problem with pitch, however, was not so much its variations over geographical space as its inconsistency over time, specifically its ascent. In the words of the prominent organ builder Aristide Cavaillé-Coll, "over a period of about a century, pitch would have risen a tone, or half-a-tone per half-century, and by continuing this ascending progression we would go through all steps of the chromatic scale."[51] Allegations regarding the escalation of pitch over time and the perception that this represented a danger carried immense urgency because vocal music was of crucial importance to French audiences. After all, in France, unlike Germany, opera remained the central musical genre for most of the nineteenth century. Like Lissajous in 1855 and Marchal in 1859, many commentators had warned of the physiological consequences of rising pitch for singers' voices during the previous decades. In 1840, for instance, the influential music critic François-Joseph Fétis denounced the "murdering of singers" and demanded the state act to protect vocalists from the threat of high pitches.[52] In the same way as Lissajous, the critic put the blame for this rise on instrumental music. Composers and performers' search for "shine in sonority," he argued, had driven a constant escalation in performing pitches since the middle of the eighteenth century. Whether or not Fétis and Lissajous were right, their reasoning nonetheless reveals the connection between debates about pitch standardization and the emergence of timbre as a central category of musical aesthetics. The development of instrumental genres like the symphony in the second half of the eighteenth century brought with it an unprecedented attention to instruments' sonority that affected the way composers wrote and listeners experienced music.[53] Within this new regime of musical appreciation, pitch was a defining feature of how an instrument sounded, or its timbre. A vital ingredient of timbre was the concept of "brilliance." Brilliance epitomized the relationship between pitch and timbre. To be sure, the notion encompassed more than just the pitch and tone color of instruments made of certain materials, especially brass, referring as well to a musical work's virtuosity or eloquence.[54] Musicians and commentators, however, often used it in reference to the instruments' register. For example, the Belgian composer and theorist François-Auguste Gevaert explained that "the bugle has the brilliance of an elevated soprano," and that "on the *E-string* [i.e., the upper string], the sound of the violin is bright and dazzling."[55] In contrast, Gevaert continued, "the viola's sonority is duller, its timbre is grave and earnest," adding later that the instrument "does not have enough . . . brilliancy to perform successfully in great concert solos."[56] Gevaert's observations are exemplary of pitch's role as a defining component of tone color within mid-nineteenth-century musical aesthetics.

In the first decades of the nineteenth century, discourses on timbre became increasingly critical of what some musicians and theorists perceived as the "excesses" and "exaggeration" of contemporary orchestration.[57] In this context, high pitches became the targets of much criticism. Fétis expressed such aesthetic prejudices: his condemnation of high pitches was part of his broader denunciation of the orchestra's noisiness. For him, composers after Mozart had reached the limits of sound power, and composers in recent times had exceeded them—except for Italian masters such as Donizetti, who cared for the voice more than for the orchestra.[58] To protect the voices of singers, Fétis recommended that composers seek "new effects in a variety of means that would provide singers with time to rest, which would be precious for the conservation of their organ [i.e., voice]." Going beyond Fétis's warning, the pedagogue Gustave Bénédit even contradicted the idea that high pitches were more expressive, presenting them as the enemies of expression and, ultimately, a "source of boredom" for listeners. In the book he published in 1860 to respond to the controversies surrounding the adoption of the diapason normal, he stated:

> If the singer persists in placing himself beyond the natural range of his voice . . . he can no longer nuance, swell, or diminish the sound. Once in this place, he is no longer leading his voice, but his voice is leading him. And then comes undoubtedly monotony, the worst of all faults, for it is the mother of boredom.[59]

At a time when the performance of a composer's opera represented the summit of a career, and France's entire musical education system was organized around sustaining Paris's musical scene with fresh voices from throughout the nation, the well-being of singers was of great importance,[60] especially to France's two leading musical institutions, the Opera and the Conservatory, both under the authority of the government.

Movements for pitch reform in the aftermath of the industrial revolution almost always derived from aesthetic and cultural transformations that engendered new relationships with past music. The invention of the diapason normal both resulted from and participated in the extensive development of musical canons. In the same way that calls for uniformity across space resulted from the encounters of musicians and artifacts based in different places, the perception that pitch had changed over time was the outcome of engaging with musical works from earlier time periods. Records show efforts to fix pitch at France's Opera and Conservatory commencing as early as 1800. At the Paris Opera, the music of Christoph Willibald Gluck and other eighteenth-century composers was increasingly conscripted within a venerated canon of operatic works. Some singers complained that their voices were challenged when they had to perform at higher pitches works that had originally been written for lower standards of pitch. In 1801, Joseph Bonnet, the director of the Paris Opera, recommended the pitch in use at his institution be lowered a quarter of a tone. Despite this effort, singers continued to criticize the orchestra's high pitch. Responding to the complaints of the opera's prima donna, the state established a commission in 1824 to investigate the issue.[61] The director of the Paris Opera, François Habeneck, appointed the members of this body. It was composed of the musicians responsible for France's leading musical institutions, including the conductors of the Opera and the Théâtre-Italien, as well as the heads of the Royal Chapel and the Conservatory.[62] After studying the problem of escalating pitch, they recommended pitch be lowered by approximately three-eighths of a tone.[63]

The 1824 decision was evidence of the far-reaching aesthetic implications of pitch regulations. Since the commission feared that "in the first days of its adoption, this improvement might surprise the audience by disaccustoming them to the brighter effects to which their ears have become habituated over the past few years," they recommended that the Royal Academy's diapason become "at the same time the diapason of the lyrical theaters, the Royal School and the Chapel."[64] In other words, the first extensive scheme for regulating France's musical practices was the result of combined concerns over singers' health and listeners' aesthetic experiences: this latter problem raised broader social and financial questions relative to competition between the premier musical institutions in Paris.

The Conservatory had demonstrated a similar preoccupation with the sustainability and popularity of the French art of singing during the First Empire. Given that "for a long time, music lovers have complained that orchestras crush singing," the Conservatory argued in 1812, "the most beautiful voices exhaust themselves," and, "in order to prevail among instruments, go beyond the limits past which singing can have neither grace nor justness in expression, nor feeling."[65] Members of the Conservatory's teaching committee, believing it fell within their responsibility to protect the pool of French voices from the detrimental impact of pitch escalation on "young students whose resources nature has not yet completed and who were constantly forced to make efforts that used up their talent,"[66] lowered the standard of the orchestra accompanying the institution's students when giving public performances.[67] They also encouraged other musical institutions to follow their lead.

Early attempts to regulate pitch at the Conservatory and the Opera followed similar procedures. Far from engaging with mathematical or physical considerations, the members of the Academy of Fine Arts tasked with resolving the question proceeded by "comparing the tuning devices of all orchestras in Paris, and chose the median between those of the Imperial academy [i.e., the opera], the opera buffa, the Emperor's chapel, and [the Conservatory's] public exercises."[68] The musical authorities Habeneck had brought together for the 1824 opera commission approached the issue in a similar fashion. The opera flutist and oboist were asked to play their *A*s and the latter of these was compared with the pitches of older instruments, as well as with forks used at various times in different musical institutions. Members of the commission determined to lower the pitch by listening and comparing pitches, referring to the differences between these sounds by using musical systems of scales, tones, and subdivisions, instead of any mathematical or experimental theory.[69] The work of both the 1812 Paris Conservatory and 1824 Paris Opera commissions was thus quite detached from any acoustical considerations or technologies: these were very much musical, rather than scientific, investigations. In determining the pitch of historical tuning devices and instruments, the composers and conductors used their ears and made reference exclusively to musical concepts. The instruments central to the physical sciences had little influence in the Conservatory where their inquiries took place. What is more, it was not long before the 1824 reform was undone and the "brighter" sounds of the orchestra returned to the Opera stage. According to the musicologist Adrien de La Fage, "when Rossini finally introduced his works on the Opera stage [in 1826], the pitch he found was so low that it took away from the instruments their brightness and their vigour. The reasons of an artist whose authority had such great weight prevailed, and the pitch was once again raised."[70]

These early schemes of pitch regulation show how, at the beginning of the nineteenth century, pitch came to crystallize the tension between two contradictory trends: the canonization of vocal repertoires, requiring a relative stability in performing pitches on the one hand, and a new taste for high sounds resulting from the development of orchestral music on the other. To be sure, canonization was not a new phenomenon: since the death of Louis XIV, the Royal Chapel at Versailles had perpetuated the performance of the works composed during the king's reign.[71] What was new was the conflict between the persistent presence of past vocal musical works on musical stages, and new instrumental developments that increasingly rendered higher pitch a feature of acoustic aesthetics. In this context, tuning forks became pieces of evidence attesting to these chronological transformations. As witnesses to endangered, disappearing musical worlds, they were turned into archives of sonic pasts and used to regulate the coexistence of musical works from different time periods.

Throughout the first half of the nineteenth century, musicians, acousticians, and instrument makers all made efforts to unify pitch, yet their approaches involved competing standards: most obviously, for instance, in the musicians' use of *A* (corresponding to one of the open strings of a violin) versus the use of *C* favored by physicists. The next section considers how these different notions and experiences of pitch came together under the auspices of the state. Although Lissajous created the conditions for the negotiations between France's musical and scientific communities, the commission of 1858–1859 embodied this radical transformation, bringing together the skills and expertise of musicians with the apparatus and concepts developed in the science of acoustics. In this moment, the parties' sharing of musical and acoustical knowledge and skills came under immense strain: as much as a collaboration of artistic and intellectual traditions, it was to become a tense confrontation between different interests.

MANUFACTURING THE FRENCH PITCH

In the context of a growing consensus that pitch had to be fixed, Lissajous's suggestion to the SEIN met with great enthusiasm.[72] Instrument builders were the first to formally discuss the issue. France's organization of musical labor was marked by a strict division between different types of instrument making, especially between those working with wood and craftsmen specializing in metal. Despite this, in 1856 members of the Société des facteurs de piano, the first union of instrument builders (established in 1853), invited counterparts from other domains to join an "interdisciplinary" conversation on the issue, along with other representatives from across the musical world.[73] On 9 June, the Société held the first of a series of meetings gathering piano, organ, woodwind, and brass instrument makers, as well as teachers, *musicographes* (music scholars, before the creation of the word and discipline of musicology at the end of the nineteenth century), and, of course, Lissajous himself. There were no representatives from Paris's string instrument–making industry, either because they were not invited, or because the issue was not as pressing for them, given that violins could easily be retuned without the intervention of a manufacturer. For piano-, organ-, and wind-instrument makers, however, standardization promised some distinct commercial advantages.

Prior to the event, the society had sent letters to the "main musical institutions" of Europe and was eager that the responses from this survey inform the proceedings of meetings. At the first gathering, Lissajous repeated his demonstration of pitch escalation over time with the help of his various forks.[74] Instrument makers then reported on misadventures caused by lack of a common reference across different locations. The "spiciest" anecdote came from the brass instrument builder Gustave Besson, who recalled that he had once had to cut the tubes of his instruments in order to adapt them to the higher pitch of Belgian military bands.[75] His colleague Charles Louis Triébert, a maker of woodwind instruments, similarly reported that products he sent to Italy were once sent back to him because they were not consistent with local ensembles. Finally, the piano maker Claude Montal explained that even in Paris, builders faced the same challenges: he reported that the organizers of the fashionable Concerts Musard had had to cease performances featuring both organ and piano (made by Montal) because the two Paris-made instruments could not be put "in agreement."[76] Despite the clear differences in these actors' activities and the variety of issues raised, the meeting collectively agreed on the need for a common reference. This unanimity was, the critic Adolphe Giacomelli concluded, "remarkable."[77]

The consensus in favor of a common pitch did not extend to a definition of the standard or the manner in which the problem should be resolved. François-Etienne Bodin, for instance, a rehearsal pianist at the Paris Conservatory and the author of a *Traité complet et rationnel des principes élémentaires de la musique*,[78] criticized Lissajous's proposition to align the standard with the metric system. In a letter addressed to the director of the society's journal after the meeting, he argued that musicians could not tune from the note *B*, and that a sonic series based on the metric system, involving the division and multiplication of 1,000 vibrations by 10, would have "nothing in common with our musical scale. One should not seek to assimilate things that don't have anything in common," he concluded.[79] There were also conflicting views on the way pitch unity could best be attained. As he had the previous year, Lissajous suggested persuading the government to convene a commission. But the *musicographe* Adrien de La Fage, eager to defend the prerogative of the musical profession as "the only one interested in the question," argued that if the government should be involved, it was only at a later stage, after musicians and manufacturers had settled all the artistic questions.[80]

To address these questions, the meeting's participants appointed a commission consisting of Lissajous, La Fage, Besson, Triébert, Bodin, and Giacomelli, as well as the *musicographe* Aristide Farrenc, the organ maker Alexandre Debain, and the piano maker Charles Louis Franche.[81] Although their discussions only lasted a few months, this rapprochement between makers from different fields marked a turning point in the organizational history of French music professionals. During one of the following meetings of the society, its president proudly announced that the government had granted him the authorization to rename the body the Société syndicale des fabricants de pianos et autres instruments de musique, an institutional creation that survived long after discussions over pitch had moved beyond the confines of the society.[82]

Due to the lack of surviving archival material, the society's final resolutions remain unclear. As the last report on the commission's work was issued in the society's journal, its members were still gathering evidence of tuning practices in Paris, as well as in other French and foreign cities. Although the physicist Charles Delezenne, who wrote a letter to the committee that was published in the journal, advocated the adoption of the Stuttgart pitch (*A* 440), the choice of a specific figure for the standard remained an unsettled question. Eventually, the body determined that government action would be the favorable course, despite La Fage's earlier opposition.

Besides the lobbying of instrument makers, several other factors contributed to the state's subsequent decision to appoint a commission to solve the problem of a national pitch. Importantly, the emperor was eager to support the cultivation of both French musical practice[83] and the natural sciences. In the same way that his uncle had invited Chladni to present his acoustic figure experiments in 1808,[84] in 1857 Napoleon III received Lissajous at the Tuileries.[85] The president of the SEIN, Jean-Baptiste Dumas, was very close to the emperor, which is probably why Lissajous secured the opportunity to address the society's general assembly in the first place.[86]

At first, the government's decision to take action appeared to appease France's instrument makers, but it became clear that this state action was very far from what this interest group had envisaged. The commission included two scientists, Lissajous and Despretz, but was otherwise dominated by the celebrated opera composers Fromental Halévy, Daniel-François-Esprit Auber, Ambroise Thomas, and Berlioz, all of whom were members of the Académie des Beaux-Arts. It also included four government representatives, among them General Émile Mellinet, as well as Édouard Monnais and Camille Doucet, who were respectively responsible for military bands and opera houses, these being the institutions most in need of having a standard to be implemented. Instrument builders may have played a decisive part in securing the government's attention, but they were excluded from the commission and were to take only a minimal role in the subsequent negotiations.

Instead, the commission's final report laid much of the blame for the escalation of pitch on manufacturers producing instruments capable of brighter sounds for commercial advantage.[87]

Drawing on Lissajous's universalist ambitions, the commission's report expressed a hope that France would lead the concert of nations in the standardization of pitch. It declared that:

> Music is . . . a sort of universal language. All nationalities disappear in front of musical writing, for a unique notation is enough for all people, for signs, the same everywhere, represent the sounds . . . Is it not desirable that a uniform and now fixed diapason add a supreme link to this intelligent community, and that an *A*, always the same, resonating on the whole surface of the universe with the same vibrations, ease the musical relationships and make them even more harmonious?[88]

Eager for the authority of such a standard to be secured through its reference to nature, Lissajous recommended, in his 1855 lecture, the use of a *B* of 1,000 vibrations as a universal standard. This figure was not adopted by the commission, which favored an alternate basis for the standard. Despite appearing to embrace scientific rationality, the determination of the French pitch did not derive from any measure of nature. Instead, the commission decided to draw on the existing body of knowledge regarding tuning practices from across Europe. Considering that "we should start by gathering information outside and around us,"[89] the president of the commission wrote letters "everywhere where there is an opera, a large musical institution, in the cities where art is cultivated with love, with success, practiced with brio, and that one could name the 'capitals of music.'"[90] He requested details of current and past practices, as well as recommendations regarding the choice of a standard. The commission received twenty responses, accompanied by some two dozen tuning forks (see fig. 1.2). As there is no record of all the figures the commission consulted, it is impossible to know if the only people to reply were those supportive of the initiative, or whether the commission only reported the ones that shared its views. However, the standard adopted in 1859 and its accompanying report reveal much about the commission's approach to resolving the problem of pitch unification.

Lissajous and Despretz prepared two tables for presenting the data collected, containing the results of the measurement of the forks received (see fig. 1.3). In the body of the text, they presented the various opinions expressed by the authorities. Despite a consensus over the need for a fixed pitch, there

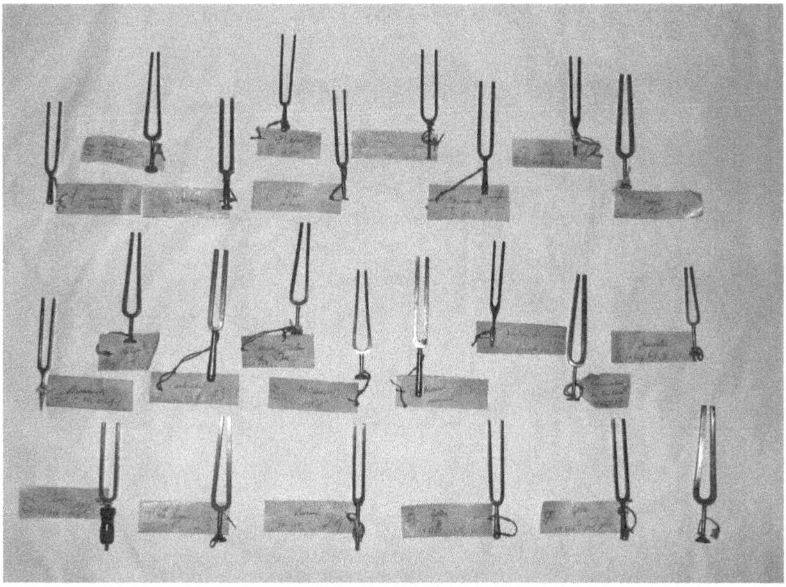

FIGURE 1.2. Tuning forks received by the 1858–1859 pitch commission. Collections Musée de la musique / Cliché Thierry Maniguet.

were different ideas about what it should be. Crucially, while three French correspondents from various *départements* recommended adopting the pitch in use at the Paris Opera, or *A* 896 vibrations, the British piano maker Henry Fowler Broadwood recommended the old London Philharmonic pitch that Despretz and Lissajous had measured at 868 vibrations. In contrast, Carlo Coccia, composer and director of the Philharmonic Academy in Turin, proposed the standard in use at his institution, producing between 889.5 and 892 vibrations.[91] The difference in pitch between the forks measured was not more than a semitone, the upper limit of which was the pitch in use in the Belgium military band, and the lowest that of the Toulouse music school. The commission connected this sonic geography to regional variations in the development of instrumental and vocal music. Reflecting the commission's preconception that instrumental music was responsible for the escalation of pitch, the report asserted that

> France counts at its two ends one of the highest pitches, that of Lille, and one of the lowest, that of Toulouse's school. One can follow, on the map, the route taken by pitch: it rises and falls with latitude. From Paris to Lille, it goes up; from Paris to Toulouse, it goes down. We see the North submitted

— 31 —

TABLEAU A.

Tableau des diapasons usités dans les principales villes de France et dans divers pays d'Europe, d'après les types reçus par le ministère d'État.

ORIGINE.	NOMBRE de vibrations par seconde.	DISTANCES au diapason de l'Opéra de Paris, mesurées en vibrations.	DISTANCES au diapason de l'Opéra de Paris, mesurées en fractions de ton moyen.
FRANCE.			
Lille	904	+ 8,0	+ 0,077
Paris — Grand Opéra	896	"	"
Paris — Théâtre Italien	896	"	"
Marseille	894	— 2,0	— 0,019
Bordeaux	886	— 10,0	— 0,096
Toulouse — Théâtre	885	— 11,0	— 0,106
Toulouse — Conservatoire	874	— 22,0	— 0,210
PAYS ÉTRANGERS.			
Bruxelles. (Musique des guides.)	911	+ 15,0	+ 0,144
Londres — N° 3	910,4	+ 14,4	+ 0,138
Londres — N° 2	905	+ 9,0	+ 0,087
Berlin	903,5	+ 7,5	+ 0,072
Saint-Pétersbourg	903	+ 7,0	+ 0,067
Prague	899,5	+ 3,5	+ 0,034
Leipsick	897,5	+ 1,5	+ 0,014
Munich	896,2	+ 0,2	+ 0,002
La Haye	892,3	— 3,7	— 0,035
Pesth	892	— 4,0	— 0,038
Turin / Wurtemberg / Weymar	889,5	— 6,5	— 0,062
Brunswick	887	— 9,0	— 0,086
Gotha	886,5	— 9,5	— 0,091
Stuttgard	886	— 10,0	— 0,096
Dresde	882	— 14,0	— 0,134
Carlsruhe	870	— 26,0	— 0,250
Londres. (N° 1.)	868	— 28,0	— 0,269

FIGURE 1.3. "Tableau des diapasons usités dans les principales villes de France et dans divers pays d'Europe, d'après les types reçus par le ministère d'État." Rapport et Arrêtés pour l'établissement en France d'un diapason musical uniforme, 32. Archives nationales de France.

eminently to the contact and predominance of instrumental art, while the South remains faithful to the uses and the good traditions of vocal studies.[92]

This extensive collection of data allowed the commission to produce a geography of pitch that substantiated its proposed measures to protect vocal music from the detrimental influence of instrumental music. Drawing on evidence boastfully described as having been "measured with all the precision of science,"[93] the commission continued its investigations into existing musical customs: "It was obvious that the greatest lowering possible was of half a tone," the report observed, and "that a greater discrepancy was neither practical nor necessary; and on this point, the commission was unanimous." The idea of a full semitone "encountered some adversaries,"[94] the report continued, and the commissioners were thus confronted with three options: lowering of either a semitone, a quarter of a tone, or less than a quarter of a tone. Despite the use of vibration as a unit to measure and present the data collected, when it came to arbitrating between these different musical practices, it was the musicians' rather than the acousticians' tools that seemed most relevant to reducing pitch values. Moreover, in determining the standard the committee ascribed greater authority to musical knowledge than to all other fields. As instrumental music was believed to be detrimental to contemporary musical practices, the commission invoked the authority of the voice in its recommendation of a unifying pitch. The report claimed that

> the composer has in his head, in his imagination, one could say in his heart, the natural type of voices. The phrase he writes is dictated by a singer whom he alone can hear, and who always sings well. That singer's voice—flexible, pure, intelligent, and in tune—is fixed by a true and moderate pitch inhabiting the ear of the composer.[95]

This romanticized image portraying the composer's creative process as subjective, individual, and internal also happened to embody the sociopolitical structure of France's musical system, in which opera composers were the ultimate authorities. After some discussions, the quarter tone secured the most votes. The commission believed that this would bring

> a sensible moderation to the studies and the work of singers, without causing a great disturbance in habits; it would sneak, so to say, *incognito*, into the presence of the public; it would render the performance of old masterpieces easier, it would bring us back to the pitch used about thirty years ago, to

the time of the production of works that have mostly remained part of the canon, and they would find again their prime conditions for composition and performance.[96]

This justification for the commission's choice is illuminating, as it builds upon earlier arguments concerning the health of singers and the preservation of masterpieces. But more specifically, whereas Lissajous referred to the reign of Louis XIV, and Habeneck prioritized sound objects from the time of Gluck, the pitch endorsed by the commission was clearly intended to revive the golden age of grand opera and its 1830s masterpieces such as *Guillaume Tell*, *Robert le Diable*, and *La Juive*. Just as opera houses and concert halls were perpetuating this canon,[97] the commission's report resolved that pitch should be kept at the level it had been during this glorious period of France's musical history. The commission arbitrarily set this level at 870 (435 hertz), a quarter tone lower than the pitch in use at the Paris Opera in 1859 (*A* 896, or 448 hertz). The commission's president, Halévy, was the author of several successful *grands opéras*, the musical genre most closely associated with the representation of political power in nineteenth-century France.[98] In choosing *A* 870, Halévy and his peers were deliberately protecting the practitioners of this genre, vocalists, from the assaults of new instrumental music. The state would be the guardian of this standard, and thereby secure the grandeur of French music. In turn, French musical genres would ensure the nation's cultural influence at an international level.

Through the choice of such a low standard, the commission reaffirmed the superiority of traditional vocal repertoires over newer instrumental genres. *A* 870 was a means of maintaining a musical order inherited from the ancien régime. In reference to the perceived escalation of pitch, the commission asserted that "religious music, dramatic music, suffer from this movement without being able to defend themselves from it, or seeking to escape from it."[99] To try to rescue these genres and their performers, the diapason normal at *A* 870 was to be adopted by "all musical institutions . . . authorized by the state."[100]

Following Montal's vision and Lissajous's recommendation, the commission's work had been divided between scientists and musicians, the former being charged with objectifying tuning practices across Europe, and the latter empowered to select the level of the nation's pitch. In this way the natural sciences were relegated to the technical role of assisting the decision of musicians—fundamentally composers who reiterated the same choice and logic of their predecessors on the 1824 commission. Lissajous and Despretz may not have had actual authority over musical practices, but their work provided a seal of authority for the musicians. Ultimately, however, musical pitch does not resemble other standards, and the French diapason normal was based not on the laws of nature nor on the reality of manufacturing practices, but was instead grounded in the history of French opera. As a result, its authority was far from certain.

Furthermore, unsurprisingly, France's standard pitch failed to produce a consensus among the actors involved in the debates. Both Lissajous and Berlioz had warned that choosing a low standard would make its implementation difficult. The latter instead advised the adoption of the Paris Opera's pitch of *A* 896.[101] Aristide Cavaillé-Coll preferred an intermediary solution: based on the measurements made since the beginning of the nineteenth century, he calculated an average pitch of 888 vibrations, equidistant from the Stuttgart and the Paris Opera's pitches. The organ maker argued that this standard would produce full numbers and have the great advantage "of conciliating the demands of the physical science and the needs of the musical art." It would

> consecrate modern tonality and end the notable difference between physicists' and musicians' pitch. This rapprochement of art and science would, for us, be the safest guarantee for the adoption of the measure and the conservation of the nineteenth-century pitch.[102]

In stark contrast, Giuseppe Staffa, director of the Teatro del Fondo in Naples, and publisher of the local acoustic-focused journal *La Musica*, went so far as to reject the very idea of pitch uniformity. La Fage relayed Staffa's remark to French audiences that "a unique pitch cannot suffice for all musical uses,"[103] as well as his suggestion to adopt four different pitches, corresponding to the traditional taxonomy of places and musical uses: chamber, theater, church, and army.

A THEORETICAL STANDARD?

Disagreements over the theoretical definition of the standard were only part of the problem. From the rooms of the Paris Conservatory where the commission had met to those of musical institutions throughout the country, it was a long way from the creation of this abstract norm to its implementation on a national and, as the standard's promoters hoped, international level. The standardization of pitch was not merely a question of replicating identical tuning devices, all consistent with a single national standard fork, although this material process was in itself a huge challenge. Rather, implementing a new *A* also involved adapting existing instruments and producing new ones, as well as fashioning new manufacturing, musical, and listening techniques. Previous studies of standards have emphasized the resistance such unifying objects encountered and the importance of examining the mechanisms that enabled their adoption. For example, Bruce Hunt and Simon Schaffer have shown that while, by 1861, British physicists and telegraph engineers had produced increasingly accurate electrical measures, it was not until 1881 and the International Electrical Congress held in Paris that a multilateral agreement was reached over the adoption of the ohm, volt, farad, coulomb, and ampere.[104] And, as Ken Alder has argued, it took over a hundred years for the metric system to become widely adopted in France, in spite of the emergence of a strong consensus around it during the French Revolution.[105]

In calling for the adoption of a standard pitch in 1854, Delezenne had warned that "in adopting a pitch . . . one will have done nothing if necessary measures are not taken to guarantee the conservation of this pitch and, above all, its identity in all French orchestras."[106] Introducing sonic uniformity across the nation, however, proved to be immensely difficult. With the exception of the Paris Opera, the state did not provide musical institutions with the financial means to implement the decision. Changing pitch was expensive, as orchestras had to either buy or adapt all of their wind instruments. While the government made significant efforts in Paris, it did not provide departmental administrations with sufficient means to implement the new standard beyond the capital. The one exception to this general lack of influence was the army.

In the 1859 report's conclusion, the commissioners recommended that the first step the Ministry of State should take was "that a diapason type producing 870 vibrations per second at a temperature of 15 degrees centigrade be built under the supervision of skilled men designated by [His] Excellency."[107] In the same way that a standard meter had been deposited in the legislative chamber in 1799 as a reference for the metric system, the *arrêté* of 16 February stated that a model of the diapason normal be stored at the Paris Conservatory.[108] Here, too, controversy arose over the way in which the standard should be physically embodied. Throughout the first half of the nineteenth century, acousticians had a number of different technologies. Whereas Chladni considered the metal blade to be the best candidate for such a prototype, Weber advocated the use of a reed pipe or a monochord, and both Scheibler and Marloye recommended the tuning fork. These diverse material solutions continued to be discussed throughout the 1858–1859 commission. While Berlioz suggested placing an organ pipe tuned to the national standard in every concert hall and opera across France,[109] Cavaillé-Coll challenged this idea by claiming that organ pipes lacked stability in comparison with tuning forks.[110] Combining ancient Chinese practices with Weber's idea, La Fage suggested a prototype relating the sonic prototype to the metric system, namely using a tube of one meter from which the scale would be deduced.[111]

The minister, unsurprisingly following the recommendation of the commission, asked Lissajous to take charge of the delicate task of producing the initial prototype. In 1855, as part of his efforts to build an apparatus for his spectacular demonstrations of acoustic curves, Lissajous had joined forces with the instrument makers Lerebours and Secrétan. Now, together, they produced not one but two prototypes, both held at the Paris Conservatory and today in the archives of the Musée de la Musique. Invoking the precedents of the platinum meter and the gold kilogram, the first standard tuning fork was made of the same precious, allegedly unalterable metal: gold (see fig. 1.4). Not intended for daily consultation, this golden fork was accompanied by a second prototype made of wood and steel. The cheaper fork would be used as the model on which future instruments were to be based. The Paris-based instrument maker Rudolph Koenig would subsequently manufacture the finest-quality forks after having established his authority by replicating the number of vibrations of Lissajous and Secrétan's prototype.[112]

In order to enforce the new pitch, the French government also required that "all musical institutions authorized by the state will have to be equipped with a fork verified and stamped, in conformity with this official prototype." This proclamation was followed by a second *arrêté*, issued in May, detailing the conditions of the standard's implementation. Lissajous was put in charge of verifying all forks at a Bureau des diapasons at the Paris Conservatory, where the prototype was stored.[113] At first this office seems to have successfully contributed to the tuning of the nation. Archival material suggests that during its first year of operation, Lissajous checked more than 1,500 tuning forks.[114] The history of the *bureau*, however, is one of a gradual decline. In 1872, Lissajous was asked to deliver acoustic lectures to justify his salary, given the lack of activity in his regulatory position; and Lissajous's successor, Sandoz, saw his salary reduced by half, a prelude to the office's total abolition. In a letter to the Ministry of Fine Arts of 1898, the director of the Conservatory claimed that "no fork ha[d] been submitted" to the Bureau des diapasons during the previous three years.[115] In 1917, the new director of the Conservatory affirmed that the institution "did not need the services" of Lissajous's successor as inspector, since "no fork had been submitted" over the course of that year.[116]

FIGURE 1.4. Diapason, Jules-Antoine Lissajous, France, 1859, E.378. Collections Musée de la musique / Cliché Jean-Marc Anglès.

The decadence of the tuning forks' office epitomizes the failure of the French state to enforce its new standard. Even at the Paris Opera, the adoption of the diapason normal proved troublesome, despite this supposedly being the model that other musical institutions across France were to follow. Initially, there were uncertainties over the best way of retuning the Opera's orchestra. The question was not so pressing for violins and other string instruments, which only required slight adjustments.[117] But the situation was more complex for wind instruments, especially for those with lateral holes. Although it is possible to change the tuning pitch of a flute, clarinet, or oboe by changing the size of its pipe, the instrument's intonation becomes faulty if the distance between the holes remains unchanged.

Some instrument manufacturers proposed to meet the new standard through minor adjustments, but others, because they either considered this solution to be acoustically unsatisfying or saw the new standard as a commercial opportunity, argued that the musicians should buy brand-new flutes, clarinets, oboes, and bassoons.[118] A second difficulty was the refusal of the Opera musicians to pay the expenses of introducing the new pitch. Instead they sent a collective letter to the orchestra director, requesting the Paris Opera's administrators to provide them with new instruments.[119]

The Opera had little choice other than to accept the instrumentalists' demands, but nevertheless refused to pay, instead asking the government to establish a special fund to cover the expense. To hasten the process and ensure that the Opera's struggles would not affect the standard's introduction beyond Paris, the government acceded to the request. The total bill for retuning the Opera's instruments came to 7,049 francs,[120] corresponding to about half the price of the most expensive organs in Paris at the time. This financial intervention, however, would not be repeated in other contexts.

All in all, the process of implementing the new pitch at the Opera reveals two insurmountable problems. First, the musicians exerted considerable power in the negotiations, and the state had little option but to submit to their demands. However definitive the national measure appeared, the fact that its implementation relied on the will and cooperation of performers effectively limited the government's authority. Second, there was a limit to the government's willingness to spend a large sum.

Other Parisian institutions seem to have followed the Opera's example, but the situation was much more complex in the *départements*. In response to the confusion of local administrations over how they should introduce the measure, the government circulated a letter to all prefects on 21 January 1860.[121] Offering a rough estimation of the cost engendered by implementing the new pitch for an average orchestra, this text did not stipulate who should pay for the replacement or adjustment of instruments. As a result, the responses of *départements* were mixed. In the absence of financial support, some local administrations simply refused to comply with the state's demand. In 1862, for instance, the *Préfet du Nord* reported to the government that in the northern cities of Dunkerque and Valenciennes, orchestras were still using the same instruments and standard as before the 1859 decree, since both musicians and mayors had refused to take responsibility for the expense.[122] In contrast, the cities of Lille and Lyon provided their operas with the means to renew their stock of instruments, effectively emulating Paris. Between these extremes was a range of intermediary responses from France's leading cities, including Toulouse, Marseille, and Bordeaux.[123]

If state-funded institutions sometimes rebelled against governmental authority, the myriad independent musical societies of the country offered even greater examples of indiscipline. To understand their reaction, one must recall that they were primarily composed of brass instruments, which meant adopting the diapason normal would involve replacing every single instrument. There is little archival material surrounding the activities of these institutions. However, where records of societies do exist, there is little within them to suggest that they adopted the new pitch, given that the financial effort required would certainly have left evidence. Furthermore, during the 1867 World's Fair held in Paris, members from these societies took part in shared performances. Evidence of the lack of a uniform national pitch was the organizers' decision to divide society musicians into two groups, one for those who played to "old pitch" and one for those using the "new pitch." An overwhelming number of these performers were placed within the first group, suggesting that even seven years after the diapason normal's introduction, it had very little impact on the myriad musical societies across the country.[124]

One exceptionally well-documented case, however, shows that in some instances, these local societies made great efforts to comply with the state's demand. The Société musicale de Caen renewed all of its instruments for about the same amount of money that the government had spent on retuning the Opera orchestra. This example, however, had unexpected consequences: the society sold its old instruments to other users, thus flooding local markets with instruments tuned to their old pitch.[125] Over time, resistance to the reform from many musical institutions actually increased the number of pitches in use by creating a new market for instruments tuned to *A* 870.

The one domain in which the standard was successfully introduced was the army, as part of broader efforts to standardize military practices. An imperial decree of 26 March 1860 imposed the use of the new norm in military bands. Drawing on practices of regular reforms and structures of control inherent to the institution, the administration took advantage of this reform to implement further changes in its bands. General Mellinet, a member of the 1859 commission, was eager to ensure the standardization of two dimensions of musical practice in his regiments. First, he wanted to standardize the number of performers comprising each band. Second, he was keen to unify the instruments the army used. Crucially, while the bands used the products of various makers, Mellinet wanted to see Adolphe Sax appointed the sole provider of military instruments. In addition to this support, the army possessed the means to regulate the practices of its orchestras and brass bands; such means were clearly missing in the rest of the country.[126]

CONCLUSION

As in the history of other measures, the creation of the diapason normal reveals the intersections between commerce, state intervention, and scientific knowledge. But the fact that this was the regulation of an art form complicated the story. The history of *A* reveals the limits of scientific and political authority over artistic questions. To some extent, the standard was not "French" but "Parisian," and its adoption mirrored the social, political, and aesthetic hierarchies that characterized the country's musical life. Despite its weaknesses, however, the French pitch proved remarkably appealing to international audiences. France's apparent ability to govern its arts made a great impression on other nations that lacked the centralized power of Napoleon's regime, offering a precedent that other countries were inspired to emulate. At the same time, the world's increasing interconnectedness provided favorable conditions for the circulation of the standard. In the following chapters, I explore how this French standard pitch traveled abroad, triggering both sonic unification and diversification.

CHAPTER TWO

Sounding the World

Nationalism, Internationalism, and the Travels of the French Pitch

> The new pitch, although it does not concern the spiritual, but the technical part of art, remains after all a chapter of cultural history. From right and left, and from Paris to Petersburg, it has expanded its threads and will soon serve as a musical railway network, to connect all cities of culture.
>
> EDUARD HANSLICK

The French pitch, as deeply embedded as it was in France's political and musical cultures, seemed to hold little promise for a bright international future. Yet by World War I, it resonated in all corners of the earth. How did this happen? And what do the travels of the French pitch tell us about the role of sound in the making of a globalized world in the second half of the nineteenth century?

From the mid-nineteenth century, the globalization of music made the lack of a unified pitch seem increasingly detrimental to the movement of musicians and musical instruments. To be sure, worldwide interconnectedness was not a new phenomenon, and as mentioned in the previous chapter, at the beginning of the seventeenth century the scholar Marin Mersenne had already imagined a means that would enable "all the musicians in the world," from "Paris to Constantinople, in Persia, in China, and elsewhere," to perform a musical work "in the tone it needs to be sung."[1] Composers', performers', and instruments' lives had long spanned multiple borders. Why, then, did the lack of consistency in performing pitches in different locations only come to be seen as such a problem in the nineteenth century?

To begin with, France, in creating an apparently rational solution, had set an important precedent for other nations, one that seemed to indicate that it was possible for a nation to unify pitch—and, thereby, protect its musical canon. What is more, in the last decades of the nineteenth century, more musicians were traveling more often to more places than ever before. In this context, the lack of a unified pitch required that wind players carry multiple instruments with them tuned to various pitches and that singers adjust to a broad variety of standards, many of which were vocally challenging higher pitches. In addition, the world's increasingly connected commercial markets, combined with industrialization, promised new business avenues to instrument builders. In the absence of pitch uniformity, however, it was difficult for manufacturers to take advantage of these economic opportunities.

Until the nineteenth century, traveling musicians adjusted the pitch of their instruments by using their ears, lips, and hands. But in their attempt to create more precise and stable technologies, industrial instrument makers severely limited performers' ability to control tuning and adapt it to various musical contexts. At the same time, efforts to increase precision in tuning focused attention on pitch accuracy, which increasingly became synonymous with quality of musical execution, even if this was an elusive process and, therefore, a difficult one to document. With their emphasis on solfège and exact intonation, the Paris Conservatory's curricula and examination procedures, which served as the models for new institutions of musical education across the world, offer a good example of this evolution. During the early modern period, religious institutions had been the centers for musical education. There, moral instruction and knowledge of liturgical uses were at the center of singers' training.[2] In contrast, the Conservatory, founded in 1795, promoted an agenda underpinned by ideas of scientific precision and rationality. Under this new regime, the ability to sing or play in tune was one of the central criteria for the evaluation of pupils.[3]

While the creation of the diapason normal was of interest to other nations, it took place at a time of great tension between France and its neighbors. Seeking legitimacy through military power, and in an attempt to emulate the accomplishments of his militaristic uncle Napoleon Bonaparte, the emperor Napoleon III went to war with Russia (1853–1856), Austria (1859), and Prussia (1870), while France's relations with Britain were also continually strained.[4] Adding to this tense international context were political anxieties over the autocratic procedures employed to implement the diapason normal. The making of a national pitch in France owed much to the existence of a highly centralized state, which most European countries then lacked. The question for the nations interested in emulating France was how pitch could be unified without the governmental apparatus of the Napoleonic state.

The problem elicited a broad variety of responses from across Europe. For instance, in 1860, the British Society of Arts tried to introduce the French pitch throughout Britain. Determined to approach the problem in a more liberal fashion than the French, however, the members of the Society of Arts' pitch committee, appointed to study the question, instigated a nationwide survey, which resulted in the adoption of a distinct measure: a *C* of 528 vibrations.

Despite individual countries' agreements and disagreements over the diapason normal, a shift was occurring in imaginations of the standard—away from the national and local, and toward the global. International tensions and differences between political systems might be shaping a cacophony of rival standards, but the same forces contributed to an internationalization of pitch debates. The reception of the French pitch across Europe was part of a complex game of alliances and counteralliances that produced some unexpected outcomes. Crucial to the diffusion of the measure were the efforts of Austria. In 1862–1863, the country engaged in a series of bilateral negotiations to introduce the standard in German states. Austria's commitment to promote a standard created by one of its most trenchant adversaries was less a reflection of the country's relationship with Napoleon III's regime than an attempt to assert the increasingly fragile authority of the Habsburg monarchy over the German Confederation. It was also part of the Empire's commitment to unite its various ethnicities through cross-cultural standards. In 1885, Austria further orchestrated the adoption of the diapason normal by eight European countries during an international conference held in Vienna as an exercise of soft power amid the continuing decline of its influence in Europe.[5]

The French commission envisioned the diapason normal as a universal measure. In truth, it was embedded in Paris's contemporary operatic culture at the time, a fact that undermined both its national and international authority. Thus, European discussions about the French measure first proceeded alongside each other at the national level, without much interaction. But unification soon became an increasingly international question. Pitch reforms were part of wider processes of globalization, including the worldwide spread of news, the touring of virtuosos, military expansion, religious missions, and the dissemination of Western musical technologies.

The history of the French pitch's travels broadens previous understandings of standardization processes. The creation of standards for electricity, railway, and telegraphy required the uneasy collaboration of physicists, engineers, and various state authorities, but the unification of musical frequencies hinged on cooperation between much more diverse social groups, from sovereigns to musical amateurs such as parish singers, wind band players, or piano owners. The formation of a musical standard was very subjective, and involved questions of taste and aesthetics. In this sense, musical pitch was arguably the most humanly contingent of all nineteenth-century measures.

What is more, unlike other unifying systems, pitch was not a material object, but an elusive phenomenon being constantly redefined by ever-shifting assemblages of human and nonhuman actors. Crucially, whenever the debates focused on questions of musical performance, the impact of temperature on tuning became problematic. Not only were musical venues susceptible to variations in heat, but temperature affected musical instruments in diverse ways. Imperialism made these problems all the more urgent, as "extreme" climatic conditions beyond Europe magnified the fundamental problem.

A DISSONANT CONCERT OF EUROPE

The work of the 1858–1859 French pitch commission attracted considerable attention across Europe even while its inquiry was ongoing. For instance, in October 1858, the *Süddeutsche Musik-Zeitung* emphasized the value of this initiative, calling attention to its "cosmopolitan" dimension. For, as the journalist wrote, "the discussion of the possibility of a reduction of modern pitch is a question that interests the musical world as a whole."[6] Similarly, in August 1858, on learning that Napoleon had established a commission to determine a "uniform diapason," *The Spectator* in Britain hoped that the country would soon follow this act of "perfect national unity." Blaming musicians for raising pitch in an attempt to produce increasingly "brilliant" musical performances, the journal was confident that the only way to prevent further escalation was for the government to take action.[7]

Although it attracted some interest and admiration in Europe, the diapason normal that emerged also triggered much criticism. Conversations about pitch unification in the various parts of Europe were imbued with local sociopolitical and cultural values. To some, the diapason normal seemed unsatisfactory from a theoretical point of view because of its lack of reference to nature. Others invoked the musical past to denounce the French pitch's inconsistency with the work of historical composers. Still others raised concerns over the level of the pitch, which some considered to be too high for singers' voices. In addition to the scientific and historical character of the standard, European audiences expressed political reservations regarding the process by which the French commission had made its decision. For instance, one journalist writing in the *Süddeutsche Musik-Zeitung* felt that rather than making this a national inquiry, France should have convened an international consultation. Furthermore, the government should have appointed "practical men" in addition to composers and acousticians.[8] Pointing to the authoritarian character of the French state, the British *Chamber's Journal* observed ironically that "the French like to have things done for them by their government, even to the tuning of fiddles."[9] As a result the spread of the French pitch internationally, instead of enhancing uniformity in the years following 1859, in fact increased Europe's sonic diversity. These political points become very clear when comparing British, Austrian, and German efforts to implement the French standard.[10]

The first country to emulate the efforts of the French government was Britain, which had been home to conversations on musical pitch since the early days of the Philharmonic Society, created in 1813.[11] In 1824, this society fixed the pitch at *A* 433, later raising this to *A* 455 in 1845, attesting to the taste for high pitches across several European institutions.[12] As in France, British interest in pitch standardization resulted from a combination of scientific ambition to give music a rational basis, aesthetic concern over the conservation of past musical repertoires, and commercial considerations regarding the constitution of a national market for instrument makers. But initial efforts to introduce the diapason normal in Britain came at a moment in which Anglo-French relations were unusually strained. Throughout the 1850s there were several invasion panics in Britain, and British audiences were ever critical of Napoleon III's authoritarian manner of government.[13] This political context ultimately shaped a redefinition of the diapason normal.

The British Society of Arts took the initiative in the adoption of a standard pitch. Founded in 1754 to encourage arts, manufacturing, and commerce, the society had advocated the reform of Britain's weights and measures since the Great Exhibition of 1851 and was an eager promoter of a uniform system of national education.[14] In the wake of France's decision, Harry Chester, vice president of the society, encouraged Charles Wentworth Dilke, chairman of the institution's council, to take up the matter. While he was eager to follow the example of the diapason normal, Chester was also troubled at the political implications of making such a measure law. Although he thought a similar standard might be of value in Britain, Chester warned that this could not be implemented through the same procedures as it had been in France. As he explained to Dilke,

> in the absence of any competent musical authority legally or officially established in England, the Society of Arts might convene a conference of musical magnates, amateurs as well as professionals, composers, instrument-makers, vocalists, and instrumentalists, to discuss the subject, and to determine whether the society should frame a resolution, and get it extensively signed: to the effect that the persons signing accepted the French decision, and would use their influence to procure the adoption of the same uniform pitch. By such a measure, we might make what would be equivalent to a voluntary law for ourselves; and public opinion, thus expressed, would lead the instrument-makers generally to confine themselves to that standard.[15]

In other words, the Society of Arts would initiate the inquiry, but the British standard would be voluntarily adopted following a public consultation rather than implemented through state legislation.

Following Chester's suggestion, Dilke reported to the council of the Society of Arts in May 1859 that France had recently agreed on a standard musical pitch and that he "had consulted with many leading musical authorities . . . as to the practicality of effecting the same object in this country."[16] The Society's council summoned a preliminary meeting of scientific and musical men on 3 June 1859, to discuss "how far it would be practicable to do anything in this country in reference to it," and whether "it was desirable that one Uniform Musical Pitch should prevail."[17] Agreeing on the advantages of establishing a musical standard, the council appointed a committee charged with addressing the question.

The committee appointed to investigate standard pitch included mathematicians, musical performers, composers, instrument makers, and natural philosophers.[18] Although the appointment of such a committee recalled France's approach, Chester warned its members that they "could not in this country make a law for a compulsory uniform pitch, as had been done in a neighboring state."[19] Britain's upper and middle classes prided themselves on their liberal system of government and the laissez-faire thinking that the economy was better regulated by natural laws than by parliamentary legislation.[20] In this respect, France served as the antithesis against which British audiences defined themselves. This difference also resulted in the redefinition of the diapason standard.

Working within the liberal political framework described by Chester, the committee conducted a different kind of survey than that instigated by the French commission. Instead of consulting an international selection of tuning forks, the committee investigated national opinions. This strategy was consistent with the prominence of local government within British politics. On 28 August 1859, the committee sent letters to the leading musical societies and institutions across the British Isles, inquiring as to whether a national standard was desirable, whether it would be difficult to introduce, and at what level it should be set.

A general consensus in favor of a national standard emerged from the survey. Organ builders, piano makers, and manufacturers of wind and brass instruments were especially supportive of such a measure. Recalling previous disagreements between vocalists and instrumentalists in France, however, there were competing claims of what the standard should be. While the committee uncovered "a decided feeling, especially among violinists, in favor of a high pitch as contributing to 'increased brilliancy' in the *timbre* of the instrument," there was evidence from singers around the country that vocalists were struggling with pitch increases.[21] For example, W. Mason, conductor of Lincoln Cathedral's choir, reported that during his twelve years of teaching singers, he had observed "that three voices out of four are either broke or ruined before they are developed in consequence of the high pitch."[22] W. Lockyer, of the Vocal Association, echoed this sentiment, hoping that the committee would establish a pitch that was "attainable to the powers of the human voice," while Charles Saldman of the Musical Society of London believed "the fact to be almost unanimously acknowledged, that the musical pitch, most in use in this & in other countries, is inconveniently high . . . because it fatigues & strains voices."[23]

In light of these testimonies, it became clear that determining the number of vibrations to which the standard should be set meant arbitrating between competing claims from musical practitioners. Like the French commission, the committee prioritized the experiences of vocalists and agreed to protect the voice from the perceived damages of escalating pitch. Both the Stuttgart pitch of *A* 440 (*C* 528) and the diapason normal of *A* 435 (*C* 522) provided lower alternatives to the pitch then in use at London's Italian Opera, *C* 546, but the project of creating a measure within a liberal political framework involved mobilizing alternate forms of authority to those of the state. As a result, there was interest in selecting a pitch with scientific credentials. With William Whewell, Charles Wheatstone, Augustus de Morgan, and William Pole as members, the committee included leading representatives from the British scientific community. At the first meeting of the pitch committee, a letter from the celebrated astronomer John Herschel made the case for a pitch grounded in mathematical theory.

Herschel's argument, subsequently published in the *Leeds Mercury*, was that a standard *C* should be fixed at 512 vibrations. This value was grounded on the mathematical theory that the various octaves of a note corresponding to one vibration per second consisted of a series of powers of 2.[24] Based on this mathematical principle, *C* 512 would be the ninth octave of a fundamental note corresponding to one vibration per second. Herschel contended that this had

> a claim to universal reception on the score of intrinsic simplicity, convenience of memory, and reference to a natural unit, so strong that I am amazed at the French not having been the foremost to recognize and adopt it, when it is remembered that their boasted unit of length, the meter, is based on the subdivisions of a natural unit of space, just as the second (a universally used aliquot of the day) is of time; the one on the linear dimensions, the other on the time of rotation of the earth.[25]

Reviewing Herschel's proposal of *C* 512, the committee felt that this mathematical ideal carried a great deal of credibility: it was a pleasing "theoretical pitch."[26] Although they were not trying to compete with the diapason normal, the inquiry felt that in invoking mathematical knowledge, they were completing France's project for standardizing music. Adopting *C* 512 would not constitute an outright rejection of the French pitch, as the committee explained how the "commission recently appointed to report on the pitch in France, who appeared to have been governed by considerations of a purely practical kind (therefore ignoring mathematical convenience entirely), have decided on a pitch, certainly not identical with the pitch 512 vibrations, but differing from it only to the extent of ten vibrations per second."[27] Herschel's proposal represented only a minor corrective to the French standard.

Nevertheless, the committee felt that *C* 512, while theoretically authoritative, would be practically impossible to impose. With several orchestras using a pitch of *C* 546, a sudden reduction to *C* 512 "could not be made without great inconvenience and pecuniary loss to the body with whom the adjustment of the pitch practically rests," namely musicians, musical institutions, and instrument makers.[28] This acknowledgment that the acceptance of any standard pitch was contingent on the musical community represented a dilemma. On the one hand, the committee wanted to utilize science to invest credibility in their standard pitch. At the same time, such a standard could not secure consensus in society. A compromise was required. While the French arbitrated between different musical traditions, the British decided on *C* 528, a rough average of *C* 512 and *C* 546. The measure was positioned halfway between mathematical knowledge and contemporary musical practice. The committee thus acknowledged its limited ability to introduce a natural standard. This was made clear when information reached the committee of "considerable difficulties . . . in enforcing the new musical diapason in France, and that authority such as would never be sought for, or obtained, in this country, has found a powerful antagonism in 'the inexorable logic of facts.'"[29]

Over the next decade it would become clear just how little impact this measure would have on British musical practice. Nine years after the committee recommended *C* 528, Henry Lunn, a member of the Society of Arts' investigation, reported in the *Musical Times* that pitch was still not uniform. He argued that unless the French standard was introduced nationally, Britain would remain in a state of musical chaos.[30] Indeed, without regulation, the frequency at which Britain's premier musical institutions set their pitch continued to escalate. At the Wagner Festival of 1877, held in the Royal Albert Hall, Continental vocalists complained about being asked to sing to a pitch of *A* 455, while the celebrated Italian-French soprano Adelina Patti refused to sing at Covent Garden in 1879, asserting that the orchestra's *A* 455 was too high.[31] It would be a long time before Britain adopted a national standard of musical pitch.

Along with Britain, Austria was among the first nations to try to emulate France's creation of the diapason normal. In this country, the government took the initiative of reforming musical pitch. In 1860, the state imposed the use of the French pitch at the Opera in Vienna and, in 1862, the government launched a reform aimed at introducing the standard across the empire and the German Confederation.[32] As the administration explained, the French inquiry of 1859 had revealed that Vienna's pitch was one of the highest in Europe, suggesting that "this measure seemed all the more necessary." In addition, the foreign minister felt that the question could not be ignored by a nation "as authoritative on musical matters" as Austria.[33] Home to Mozart, Haydn, and Liszt, and the adopted country of Beethoven and many other eminent composers throughout the nineteenth century, Austria had assumed an image as one of Europe's leading musical nations. Although increasingly fostered by the development of private musical institutions, the country's musical life was a matter of state. The emperor Franz Joseph I was an enthusiastic supporter of the arts and encouraged the developments of Vienna's musical infrastructures.[34] Epitomizing the art's significance for Austrian political culture, music provided the Habsburgs with one of their favorite metaphors: "harmony."[35]

Following the 1859 Franco-Austrian War, which saw Napoleon III's armies defeat Franz Joseph I's army, the Habsburg monarchy's initiative to introduce the French pitch evidenced the improvement in relations between the two countries, favored by the arrival of Richard von Metternich as Austria's ambassador in Paris and the development of a close relationship between the diplomat and the French emperor.[36] Yet Austria's interest in pitch unification was also part of larger efforts to unify its empire and assert its international influence.[37] In orchestrating the introduction of the diapason normal across Germany, the country hoped to restore its authority within the European concert, amid political and military defeats.[38] As suggested by Hanslick's metaphor of the "musical railway network" linking Paris to Saint Petersburg (see the epigraph to this chapter), Austrian elites were eager to think of the country as a geographic center connecting Britain and France in the West with the Ottoman and Russian empires in the East.[39] Just as travel and communication infrastructures did, pitch unification would contribute to establishing Austria's centrality in a globalizing world.

More specifically, Austria's pitch reform of 1862–1863 intended to secure the country's influence within the German Confederation.[40] Since the Congress of Vienna of 1815, Austria had been eager to impose itself as the leader of the region, but its ambitions had been constantly threatened by Prussia.[41] Prussia dominated the Zollverein, a German customs union established in 1834 to spur regional commercial integration through common tariffs and economic policies and from which Austria was excluded. Moreover, in the aftermath of the 1848 revolutions, the Austrian government released a plan for the reorganization of the German Confederation. Prussia responded by creating the rival German Federation, and the two organizations fought over intervention in Hesse's domestic affairs.

Prussia's hostility to Austria was all the more apparent by 1859 with the outbreak of the Franco-Austrian War in Italy, during which Austria's Germanic rival remained neutral, contributing to France's quick victory.[42] In the absence of military and political influence, Austria sought to establish its authority over Germany through cultural and economic initiatives. In particular, the Habsburg monarchy initiated the construction of communication infrastructure to unite its empire with the German Confederation. In 1850, a German Postal Union was established under Austrian leadership, and the Austrian government similarly fostered the development of the telegraph in the region.[43]

In 1862, as the Austrian government was initiating a reform to introduce the French pitch to its empire and the German Confederation, the Austrian critic Eduard Hanslick published a series of texts promoting the virtues of the measure. In an 1864 article, he called attention to the progress that the French pitch had made across continental Europe and expressed his deep satisfaction at seeing it spread from Western to Eastern Europe and become a kind of infrastructure that promised to "connect all cities of culture."

Hanslick may have been confident that pitch unification would connect the diverse centers of the musical culture of his time, but he deplored the measure's progress in his own country. This state of affairs in Austria, he claimed, was the product of political anxieties over the French approach to pitch standardization. "Our government," he argued, "wanted to stay away from the centralized despotism of France, which has brought, by decree, the whole Empire, theaters, concerts, schools, and manufactures under the authority of a single tuning fork." While "the most important music institutes of Austria could have freely adopted the new pitch," Hanslick observed that, in fact, "there is still much to be done on the path towards the voluntary learning about and adoption of the new pitch."[44]

Hanslick's comments offer a good starting point from which to analyze international pitch negotiations during the second half of the nineteenth century. Hanslick's defense of the diapason normal reveals the influence of France's cultural concerns over the alleged escalation of pitch across Europe. Like the members of the French 1858–1859 inquiry and like many reformers, Hanslick's enthusiasm for the diapason normal had an aesthetic foundation: orchestras' use of high pitches, he explained earlier, would lead to the "ruin" of music.[45] In his 1864 article, however, the critic emphasized the importance of pitch standardization for broader processes of modernization: like most of his contemporaries, Hanslick stressed the urgency for sonic uniformity by referring to the world's growing interconnectedness. In the context of a globalizing world, the lack of a unified pitch came to be seen as detrimental to the movement of musicians and musical technologies. Pitch standardization was not merely the product of these transformations, but was itself, as Hanslick put it, a "musical railway network," a means to smooth the circulation of music.

Hanslick's observations emphasizing the contrast between Austria and other nations' approach to unification captured the tension between internationalism and nationalism underpinning the debates. Unified pitch was not only a matter of international competition, but a matter of domestic concern as well, and deeply enmeshed with processes of state-building and national integration. To Hanslick, pitch was another tool that could be used by the Austrian state to secure political integration. Yet in comparison with railways, the post, or the telegraph, musical pitch was a particularly troublesome medium through which to exert Austrian influence, for despite being a technical object, it had broad and unexpected socio-cultural ramifications. Although the Austrian government hoped that it would be easier to spread the French pitch across German-speaking lands than it had been to introduce commercial agreements or a shared unit of currency, standardizing the practices of Germany's opera stages proved to be immensely challenging.[46] The Austrian inquiry of 1862–1863 not only brought to light the great variety of opinions on this question; it also revealed the extent of Prussia's authority over the musical culture of the German states. Far from securing sonic uniformity across Europe, Austria's 1862–1863 reform contributed to the fragmentation of the Continent's musical geography.

State minister Anton von Schmerling took the initiative in Austria's musical reform. A vocal supporter of constitutional monarchy, von Schmerling oversaw much of Austria's liberalization in the 1860s.[47] In 1862, in a note to Karol Lanckoronski, the *Oberstkämmerer* of Vienna's court, he unveiled his plan to emulate the work of the 1859 French commission and implement the measure in Austria as well as throughout the German Confederation. After discussing the matter with representatives of the Opera, the ballet, and the orchestra of the Vienna court theater, Lanckoronski responded favorably to the state minister's suggestion.[48]

To deliver the reform, Austria could rely on its tight network of diplomats stationed across the German Confederation.[49] On 13 May 1862, at von Schmerling's instruction, Austria's foreign minister Bernhard von Rechberg und Rothenlöwen requested Austria's diplomatic representatives in Germany to investigate whether the local governments were "inclined to introduce the French pitch."[50] In a letter reporting on France's 1859 commission, Rechberg echoed the conclusions of its report, arguing that pitch had been rising for the last century and that, if nothing was done, it could soon become impossible for musicians to perform "the scores of the old masters."[51] In introducing a standard based on "scientific precision," the letter went on, the French had solved a considerable artistic problem.[52] Although Rechberg recognized that such a reform would involve considerable financial costs, he insisted that such expenditure was of little concern in comparison with the "longer conservation of singers" that the introduction of the diapason normal promised to secure.[53]

In addition to technical and cultural implications for Germany's individual musical stages, Rechberg's letter stressed the international significance of the reform, asserting that the adoption of the diapason normal had been "greeted with great approbation" by "the entire musical world."[54] Besides solving the problems resulting from pitch's escalation, Rechberg was convinced that the French pitch was the solution to Europe's increasingly integrated musical networks brought about by the Continent's expansion of railways and steamships. "Through the recent, wonderful proliferation of previously unknown means of communication, which have considerably reduced the distance, cost, and discomfort of travels," the minister asserted, "it will be increasingly common that Viennese musicians are invited to perform in Berlin, Paris, St. Petersburg, and London, and vice versa."[55]

Rechberg expressed great hopes that there would be "no difficulty" in introducing this reform as, throughout the inquiry, "the most celebrated musical representatives from abroad had declared themselves ready to choose the French pitch." As the minister observed, the state theaters in Cologne, Dresden, and Berlin were already close to adopting the diapason normal. Given the interest in reform of German musical institutions, it was hoped that pitch would provide an obtainable demonstration of Austria's role as cultural leader of a united German Confederation. But, far from fulfilling Rechberg's ambitions, the introduction of the reform proved to be a long and complicated process.

On receiving Rechberg's letter, Austrian diplomats stationed throughout the German states forwarded the minister's question to local political authorities, who, in turn, circulated it among the members of their region's public musical institutions. The Austrian initiative brought Austria's state minister into communication with numerous German mayors, theater directors, and orchestra and choir conductors. In the smaller states like Saxony-Anhalt or the city-states of the Hanseatic League, these conversations involved only a few people—the head of the local theater or the ducal chapel, the mayor—but in Germany's more considerable states, like Saxony and Hessen, as well as in Prussia itself, the consultation involved a much greater variety of institutions and interests. In Prussia, the local representative of the Austrian empire forwarded Rechsberg's question to Otto von Bismarck, the country's minister-president and foreign minister, as well as to the chamberlain and intendant of the music of the court, the general intendant of royal spectacles, the minister of the royal house, the minister of culture, and the minister of war.[56]

Far from the consensus that Austria's government had hoped to build, Rechberg's consultation revealed a wide variety of responses. Several German states expressed their support of the reform. Members of the court theaters in Karlsruhe and Mannheim immediately agreed to introduce the French pitch in their performances, announcing that it would be implemented during the new season, respectively in September 1862 and January

1863. The representative of the opera in Karlsruhe seemed to have little doubt that the standard would soon be in use on all major German musical scenes, thus the absolute necessity to adopt it in Karlsruhe as well. The grand ducal chapel in Weimar and the Bavarian government also responded favorably to Rechberg's inquiry.

Most states, however, did not display such enthusiasm. Just as in France and England, several German authorities expressed concerns about the financial cost of the reform. For example, in Hannover the count of Platen-Hallermund explained that the reform had been under consideration for some time, but that it had not been implemented because of its expense.[57] Economic questions were all the more pressing for those in charge of military bands, as these institutions would have to renew their entire stock of instruments. In Hessen, for example, the minister of war expressed worry about the financial consequences of the reform.[58] Although Platen-Hallermund eventually agreed to introduce the French pitch in Hannover, he ordered that all military bands keep "the old pitch" to avoid considerable expenses.[59] Along with economic concerns, several states rejected Austria's proposed musical standardization altogether. In Lübeck, the senate declared that although the pitch in use in that city was around 890–900 vibrations, it was "not too high and the singers have no difficulty with it."[60] The Saxon government also initially rejected Austria's proposition on similar grounds. Arguing that their pitch was close enough to the French diapason normal, the members of Dresden's court theater refused to engage in a costly reform that seemed to offer few benefits.[61]

The German states were mostly eager to wait to see whether the reform would be introduced in other parts of the confederation. As the members of the grand ducal theater in Karlsruhe reported, they "had already dealt seriously with this question for years but did not want to be isolated."[62] In spite of these fears, the governments of Thüringen and Saxony-Anhalt expressed an interest in reform, unanimously announcing that they would definitely commit to it, but only after larger states had done so.[63] This approach is typical of standardization processes: the appeal of standards depends greatly on how broadly they are disseminated.[64] But the reaction in the German states revealed both the extent of cultural hierarchies among these states and their collective distance from Vienna.

In declaring their intentions to postpone their decision about introducing the French pitch, German states often designated as models to follow the decision of Germany's "main musical centers." Such, for instance, was the view of the mayor of Bremen announcing in a letter the city's hesitation despite its senate's eagerness to introduce the standard.[65] Several other letters similarly invoked the authority of these "centers," designating Berlin, Dresden, and Leipzig as the musical leaders of the region.[66] In this way, while Austria had hoped that German states would follow the example of Vienna, in fact they looked for direction to other cities, including the capital of Austria's great rival, Prussia.

By the spring of 1863 it was very clear to the Austrian government that there were still considerable divisions within the region's musical networks. With the exception of Mannheim, Karlsruhe, and Weimar, the Austrian reform had been considered but not firmly accepted. However strongly Hanslick expressed his satisfaction in 1864 at seeing the pitch spread across Germany, Austria's initiative had been far from successful. In Bavaria, for instance, negotiations were interrupted from 1862 to 1865, resuming only after sixteen singers of the court opera petitioned their administration, and following the adoption of the diapason normal in Berlin, Dresden, and Mannheim. The intendant of the court theater presented this request to the government at the beginning of 1866, triggering the publication of a royal decree ordering the use of the French pitch on 31 December 1866, although this was only implemented in Munich's court theater at the beginning of the 1867–1868 season.[67] Seeking guidance on the new measure's introduction, the Bavarian state established contact with the French government through their embassy in Paris to inquire about the manufacture of standard tuning devices and order hundreds of forks from the French maker who had designed the French prototype in 1859, Lerebours and Secrétan.[68]

In 1869, as part of a new Society of Arts' inquiry into musical pitch, the British foreign secretary asked European countries for information regarding the tuning practices of their musical institutions. The German states' responses revealed the coexistence of many different pitches on both a regional and a local scale. With the exception of Bavaria, the diapason normal had not been adopted anywhere as a legal standard. Its use was, however, confirmed in a number of states and duchies, including Prussia, Württemberg, and Baden. Thus, far from contributing to uniformity in practice, the introduction of the French pitch had increased pitch diversity at the local level. For instance, in Berlin the Royal Opera employed the French pitch, but "most of the chapels have still the former high pitch." Similarly, in Württemberg, *A* 435 was in use at the Royal Chapel and Opera, but neither the army nor private theaters and concerts used this standard.[69] In other words, however graphic the description of the diapason normal as a "musical railway network," uniting all "cities of culture," its impact across France and Germany was actually very limited.

Rechberg's suggested reform had a long way to travel between the office of Austria's prime minister and Germany's regional senates, theaters, and orchestras. Far from the irresistible movement depicted by Hanslick, pitch negotiations were a tedious, drawn-out process. The example of Germany highlights the complexity of mechanisms involved in pitch unification on a local level. In contrast to Austria's top-down approach, the deliberations of German states involved lateral conversations with neighboring cities and powers, national consultations, and parallel international exchanges with French state representatives and makers. The success or failure of pitch reform across German states was thus a socially and politically contingent process. Amid rising nationalism, the Austrian model of governance, based on the union of many disparate nationalities, had become weak. The French pitch provided the declining empire with an opportunity to gain cultural influence at a time it was losing real political influence.

A 435 OR *A* 432? THE AGE OF SONIC INTERNATIONALISM

The decades following saw an intensification of discussions about standard pitch, as well as a shift from bilateral to multilateral negotiations. These evolutions culminated during the first international congress on musical pitch held in Vienna in 1885, which finally marked the adoption of the diapason normal as the first international standard for music. Breaking with the cacophony of the early 1860s, the diapason normal imposed itself increasingly on the international stage during the following decades. It became Belgium's national standard in 1877, and Spain's in 1879. It also became the official pitch of the Russian Army and the imperial theaters in Saint Petersburg, as well as the royal theater in Stockholm.[70] These decisions were the products of broader transnational musical exchanges. In Russia, for example, it was Alexei Lvov who introduced the French pitch. A composer and a violinist, as well as the master of the tsar's court and director of the imperial chapel at Saint Petersburg, Lvov's career developed between East and West. The string quartet he founded in Russia performed throughout Europe and, in turn, Lvov's Saint Petersburg salon featured Western soloists, including Robert and Clara Schumann and Hector Berlioz.[71] Lvov became aware of France's efforts to unify pitch during the 1858–1859 inquiry: the members of the commission had contacted him for information on Russia's tuning practices. In response, Lvov sent a tuning fork representing the pitch in use at the Imperial Opera and expressed his fervent support of the proposed reform, arguing that "the progressive escalation of pitch is not only detrimental to the human voice, but also to all instruments." "It is above all string instruments," the violinist continued, "which have lost a lot of their sound since we are forced, because of this elevation, to use extremely thin strings, stronger strings being unable to withstand this exaggerated tension; as a result, this sound, instead of drawing closer to the human voice, is increasingly getting away from it."[72]

The global influence of the Paris Conservatory played a critical role in the international diffusion of the French pitch. In several countries, including Belgium, Spain, and Greece, it was the directors of the national music schools, modeled on the French Conservatory, who initiated pitch reforms.[73] In all three nations, the introduction of the diapason normal contributed to larger artistic and institutional exchanges: pitch uniformization was part of broader attempts to emulate France's centralized system of musical education and aimed at implementing more "modern" and "rational" pedagogical standards.[74]

While musical practitioners increasingly favored sonic rationalization on an international scale during the 1870s, the reconfiguration of the European political order brought in its wake fresh debates surrounding the French pitch. After 1871, with the final unification of Italy and Germany—two nations with strong musical histories and considerable musical interests—France's proposed standard found itself under challenge. In 1884, Italians introduced a counter-standard that soon secured significant support across Europe: *A* 432. Although it posed a threat to the authority of the diapason normal, the success of this alternative measure and the Franco-Italian competition engendered more intense multilateral negotiations. The assertion of nationalist cultural agendas at the end of the nineteenth century brought along a dual movement of unification and diversification in the negotiations. While paying lip service to the economic imperatives of pitch unification, the 1885 adoption of the French pitch in Vienna simultaneously created an increased awareness of pitch's local and national variations.

The 1885 conference initially appeared to mark the end of decades of sonic chaos, but in truth it inaugurated a new era in pitch negotiations characterized by a heightened awareness of sound's dependence upon specific cultural and environmental conditions. Confronted with considerable evidence of the lack of implementation of pitch integration, the nations assembled in Vienna approached the problem of enforcing the standard in a more comprehensive fashion than their predecessors. Doing so led them to turn pitch back into a local phenomenon embedded in specific technologies, sociopolitical contexts, and climatic conditions. The exchanges between universal and local portrayals of the standard during this era of sonic internationalism actually paved the way for increasingly relativist approaches to sound.

In the early 1880s, amid a surge of nationwide initiatives to reform pitch, musical standardization was more and more conceived of as an international concern. Information regarding various countries' acoustical standards and efforts to unify musical practice spread widely across scholarly, musical, political, and industrial networks. During the 1870s the scholar Alexander Ellis had corresponded with dozens of instrument makers, musical antiquarians, and musicians to gather evidence on past and present tuning practices. His efforts culminated in the 1880 publication of a substantial paper entitled "On the History of Musical Pitch," which secured considerable attention when it reappeared as part of an appendix to Ellis's English translation of Hermann von Helmholtz's influential acoustic treatise *On the Sensation of Tone*.[75] One of Germany's leading scientific authorities, Helmholtz had reshaped previous understandings of sound and music by emphasizing the physiological implications of hearing.[76] With his translation of this work, Ellis greatly contributed to the dissemination of Helmholtz's ideas beyond the German-speaking world, while spreading his own ideas about sound and music.[77]

Still a reference for today's performance practice studies, Ellis's work represented the first attempt to record musical pitch on an extensive historical and geographic scale, including data from as far back as the sixteenth century. Ellis's work added a fresh impetus to international negotiations (fig. 2.1). Its dual, universalist and local, spirit also manifested itself in the development of pitch debates. Specifically, it underpinned the emergence of a counter-standard to the French that still remains concert pitch's primary contender today: *A* 432.[78] Just like Ellis's enterprise was marked by a competing move toward both global and local definitions of the standard, the emergence of *A* 432 in the debates was the product of concurrent nationalist and internationalist agendas. The Belgian acoustician Charles Meerens had been the first to suggest the adoption of this measure during the work of a state-appointed commission that investigated the question of pitch unification in Belgium from 1873 to 1877. While only a few vibrations different from the French pitch, this measure nevertheless embodied a completely alternative epistemology. Accusing the French commission of having created considerable chaos by selecting what he thought to be an arbitrary measure, Meerens

OUTLINE HISTORY OF MUSICAL PITCH.

S	A		
0·0	370	Ideal lowest, or zero-point.	Church Pitch lowest.
0·2	374	Hospice Comtesse, 1700.	
0·3	377	Schlick low, 1511; Bedos, 1766.	
1·0	392	Euler's Clavichord, 1739.	Church Pitch low.
1·1	395	R. Smith, 1759; Roman pitch pipes, 1720	
1·2	396	De Caus, 1615; Versailles Chapelle, 1789	
1·4	403	Mersenne Spinet, 1648.	Chamber Pitch low.
1·6	407	Sauveur, 1713.	
1·7	408	Mattheson, Hamburg, 1762.	
1·7	409	Pascal Taskin, court tuner, 1783.	
2·0	415	Dresden chained fork, 1722.	European Mean Pitch for two centuries.
2·2	420	Freiberg, 1714; Seville, 1785.	
2·3	422	Mozart, 1780.	
2·3	423	Handel, 1751.	
2·4	424	Praetorius's suitable pitch, 1619; original Philharmonic, 1813.	
2·5	428	R. Harris, 1696; Opéra Comique, 1823	
2·7	433	Sir George Smart's fork, 1820-26.	Compromise Pitch.
2·8	435	French Diapason Normal, 1859.	
3·0	440	Scheibler's Stuttgart Standard, 1834.	Modern Orchestral Pitch, and *Ancient Medium Church Pitch.
3·1	442	*Bernhardt Schmidt, low, 1690.	
3·2	445	Madrid, 1858; San Carlo, Naples, 1857.	
3·2	446	Broadwood's Medium, 1849; French Opera, 1856; Griesbach's A, 1860	
3·4	449	=C 534; Griesbach's C 528, 1860.	
3·5	451	Lille Opera, 1848; British and Belgian Army, 1879.	
3·5	453	Mean Philharmonic, 1846-54.	
3·6	455	Highest Philharmonic, 1874; Broadwood, Erard, and (English) Steinway, 1879.	
3·6	456	Vienna, high, 1859.	
3·7	457	(American) Steinway, 1879.	
3·8	458	Great Franciscan Organ, Vienna, 1640	Church Pitch high.
4·0	466		
4·3	474	Tomkins, 1668; B. Schmidt, high, 1683	
4·5	481	St. Catherine's, Hamburg, 1543.	
4·8	489	St. James's, Hamburg, 1688.	
5·0	494	St. James's, Hamburg, 1879.	Church Pitch highest.
5·1	496	Rendsburg, 1668.	
5·3	504	Schlick, high, 1511; Mersenne, ton de chapelle, 1636.	
5·4	506	Halberstadt Cathedral, 1361.	
6·0	523		
7·0	554		Chamber Pitch highest.
7·3	563	Mersenne, ton de chambre, 1636.	
7·4	567	Praetorius, North German, very old.	

FIGURE 2.1. Alexander J. Ellis, "On the History of Musical Pitch," *Journal of the Royal Society of Arts* 28, no. 1424 (1880): 305. Courtesy of JSTOR.

claimed that the French decision should not be emulated, but "revised."[79] Rather than *A* 435, the acoustician believed that France should have selected what he called "the theoretical pitch," an *A* of 432 vibrations.

Echoing Herschel's recommendation of *C* 512 to the Society of Arts' 1859–1860 inquiry, Meerens's suggestion carried mathematical weight. Like *C* 512, this measure was derived from an initial *C* producing one vibration per second through a series of octaves; but instead of the British custom of using *C* as a standard, Meerens followed the Continental practice of *A*. To do so, Meerens used Pythagorean tuning, that is, a series of 3/2 ratios from *C* 512, which led him to determine *A*'s value as 864 vibrations.[80] In Meerens's view, this double mathematical basis—the principle of drawing from an initial *C* producing one vibration and the application of Pythagorean proportions for the calculation of the corresponding *A*—offered a permanent basis for the standard, since "in a remote future . . . supposing that it had been lost, one could recover the intonation of our sonic prototype." The authority of this measure was considerable, given its endorsement by generations of eminent acousticians, including the acclaimed maker of acoustic instruments Rudolph Koenig.[81] Comparing the diapason normal with the meter, Meerens asked "why musicians would want the meter, which constitutes the founding principle of the entire system of weights and measures, to be for them, exceptionally, 1.05 meters."[82]

This reform was not only theoretically pleasing, Meerens continued, but practically desirable, since *A* 432 "only differed by an imperceptible quantity" from the diapason normal. Not only would the change of standard go nearly unnoticed, it would require minimal material alterations. Meerens insisted that his proposed reform would thus be much more obtainable than the French 1859 decision to lower the pitch a quarter of a tone, which had required the replacement of all brass instruments in use in the French Army.[83]

Meerens failed to convince the Belgian government to adopt *A* 432 as a national standard. Due to the intensity of scholarly and artistic exchanges surrounding pitch unification, however, his arguments induced a response from beyond his country: in June 1881, a congress of Italian musicians held in Milan adopted Meerens's selected measure as a national standard. The composer Giuseppe Verdi adamantly promoted this pitch, considering *A* 432 preferable to the diapason normal from the point of view of "mathematical exigence." He thought that the difference between the two measures was "so small" it was "almost imperceptible to the ear."[84] In the words of Pietro Blaserna, a physics professor at the University La Sapienza and a leading Italian authority on acoustics, Meerens's standard "brought musical and laboratory practices into agreement."

The congress of Milan's decision had decisive implications for pitch negotiations. To begin with, the Italian government followed the recommendation of the nation's premier composers and, in 1884, issued a decree imposing the use of *A* 432 across the country. Following Italy's example, the members of a musical congress in Belgium chose *A* 432 as their standard that same year. In response, a group of German instrument makers and musicians petitioned Bismarck's government to choose between the French and Italian standards and impose the selected measure on a national level.[85]

Amid growing support for Meerens's measure, in April 1885 the Austrian government announced that it would hold an international conference to solve the problem of pitch standardization. Although this event risked a potential diplomatic defeat for the diapason normal, it actually facilitated its adoption on a large scale. Once more, musical networks played a key part: the Austrian state began examining the question under pressure from the Gesellschaft der Musikfreunde. The state took this opportunity to spearhead an international conference on the subject, pointing out that "unilateral regulation circumscribed to Austria" would only "add to the confusion and the incongruity" characteristic of Europe's musical life.[86] As in 1862, Austria was again imposing itself as the arbiter of international musical relations. In doing so, it looked to mobilize the question of pitch as a political tool to restore its waning authority within the European community, something the Italian and German unification had immensely reduced.

To secure the participation of countries, the Austrian government emphasized the conference's consultative character: once a consensus was reached, each nation would have the prerogative to decide what to do in order to implement the standard within its territory.[87] Although France and Britain, the two major European powers at the time, were absent, Austria managed to gather representatives from five countries: Italy, Hungary, Sweden, Russia, and Germany, including delegates from Prussia, Saxony, and Württemberg.[88] Like the French and British inquiries of the late 1850s, the conference offered an arena for interdisciplinary exchanges, comprising members of the main musical institutions of the various nations, from theaters and concert societies to military bands, and music critics, professors, and scientists specializing in the study of sound.

Held in Vienna from 16 to 19 November 1885, the conference was intended to examine two key questions: first, "adoption of a uniform normal pitch";

and second, "the measures to take to avoid hereafter any variation of the fundamental note adopted."[89] Austria had already resolved, prior to this event, to adopt the French standard *A* 435. As this standard was already in use in most of the countries represented in Vienna, it came as little surprise that the delegates confirmed it as the agreed-upon measure. Members of the Italian delegation acknowledged that their objective in taking part in this multilateral negotiation was not so much to promote their national measure, as giving Italian musicians the impression that their opinion mattered, that their voice had been heard on the international stage. Given the close proximity between *A* 432 and *A* 435, the difference between the two standards was largely symbolic.

The Italian delegation had an opportunity to express its views on the first day. The Austrian minister of education and culture, Carl Zeller, opened the debates by asserting that "the practical musician can become friends with any *A* that the singer can sing easily and with effect and the instrumentalist can play with sonority." In response, the Italian physicist Pietro Blaserna objected that pitch standardization was not "a mere musical question, but also a theoretical one," but added that "Italy would have no objection to adopting another pitch," providing that it be unanimously adopted or had "at least a strong majority in its favor." This won Blaserna "loud applause" from the other delegates.[90]

The second Italian representative, the conductor and librettist Arrigo Boito, claimed that "the rays of the sun of science heat and pierce all disciplines of human knowledge and the arts," and that "choosing 870 vibrations instead of the scientific standard would be an anachronism." In doing so, Boito was following the recommendation of his friend and colleague Verdi, who had written to him a few days before the conference, declaring: "*Principal aim, the standard of concert pitch.* Give in, if it cannot be avoided; but not without declaring openly, loudly, and publicly, the error, from a scientific point of view, of 870 vibrations."[91] In response to Boito's intervention, the University of Vienna's professor of physics, Josef Stefan, replied that "The number 432 has the advantage of the mathematical aesthetic. But one has to consider it in light of the fact that in France, Belgium, and Russia, the French pitch has already been introduced, so [this change] would represent a great cost for German institutions as well as for Austria."[92]

Unsurprisingly, the conference followed Stefan's recommendation and unanimously voted in favor of the French pitch. Over the following days, the delegates proceeded to discuss implementation. The choice of the diapason normal may have appeared a scientific defeat, but as Myles Jackson has emphasized, the subsequent conference discussion revealed the unprecedented authority of acoustics in pitch negotiations.[93] Eager to increase states' control over tuning practices, the delegates turned to acoustical knowledge to address the challenges raised by the introduction of the measure. However, science brought a whole new series of problems to the fore, stumbling blocks that ultimately undermined the very project of creating a unified standard of musical practice.

In preparation for the conference, the Austrian government had inquired into the ways in which France had implemented the diapason normal since 1859. In response, the director of the Paris Conservatory explained that

> the use of the diapason normal not being mandated by any law, there is no need to appoint an inspector in charge of verifying the regular use of the diapason normal; the commission mentioned in the *arrêté* of 16 Feb. 1859 never served to verify instruments; the diapason normal was only imposed on national schools, subsidized theaters, and military bands, in other words on all institutions under the authority of the state; as a consequence, instrument builders took it as a basis for making their instruments; tuning forks submitted for stamping were verified by Sandoz [Lissajous's successor].[94]

Chapter 1 showed that, with the exception of the Paris Opera, these steps had failed to secure uniformity throughout France. As a result, the head of the Paris Conservatory acknowledged that "it [was] impossible to affirm that the diapason normal [was] the only one used in France."

The resolutions passed in Vienna showed the delegates' determination to secure a wider implementation of the measure beyond what France had managed to achieve. The delegates emulated France's choice of *A* 435 as a unit, but they departed from its approach to implementation, agreeing to give countries more autonomy over musical practice. The conclusions of the conference not only fixed the ways in which nations should secure the production of standard tuning forks, as France had done in 1859, but also outlined the means by which musical institutions could ensure the introduction of the new pitch. Drawing on the observation that implementing a reform of musical pitch required much more than securing the manufacture of standard tuning forks, the delegates started more seriously examining the problem of temperature in relation to musical venues and instruments. This opened the door to countless new challenges, especially regarding the relation between pitch and musical instruments' behavior within their various environments. Although aimed at providing the conditions for universal sonic uniformity, this change of focus revealed the inseparability of pitch from local climatic conditions, ultimately marking the end of universalist approaches.

The relationship between climate and sound was not a new challenge at the time of the Vienna conference. Since the early modern era, for instance, a variety of experiments had revealed the correlation of sound's velocity with temperature and atmospheric pressure.[95] As previously shown, the diapason normal was fixed in relation to a temperature of 15°C. But as Blaserna emphasized, whereas temperature variations had only a limited impact on the pitch of tuning forks, they had a greater effect on the sound of musical instruments. Raising the temperature by 30°C would change the pitch of a tuning fork by only three-quarters of a complete vibration per second, Blaserna explained. In contrast, wind instruments and organ pipes were extremely sensitive to ambient conditions. Sound moved faster through warm air, producing higher frequencies. Although too little was yet known about this phenomenon, Blaserna cited earlier studies that had shown that a pipe that sounded at 435 at 15°C increased in pitch to 457.7 vibrations at 30°C, this being nearly a semitone higher.[96]

In addition to musical instruments' varying sensitivity to temperature variation, determining a temperature standard applicable across Europe was a challenge due to the range of different venues and climatic conditions. For instance, determining the temperature at which church organs could produce the desired pitch was impossible because of the variation in milieu between and inside these places. Acknowledging this, the report specified that organs should be tuned in relation to the specific conditions of their environments.[97] Furthermore, as Blaserna emphasized, the Vienna delegates' decision to fix the standard temperature at 24°C for wind instruments represented an oversimplification given the diversity of temperature conditions where music was presented. In the years following the conference, Blaserna further objected that "the average temperature in concerts where musical performances are ordinarily held, such as concert halls and theaters, rarely goes above 20 centigrade," since "one tends more and more to use electric lighting which as we know doesn't sensibly modify a room's temperature."[98] The problem of temperature was to increasingly define pitch negotiations after 1885.

The adoption of the diapason normal as the first international standard of pitch in 1885 involved a fundamental reconfiguration of the approach taken by French reformers in 1859. Where France's musical reform had centered on the production of standard tuning devices at the Conservatory and the activities of the Paris Opera, effectively limiting the impact of the 1859 decision, the delegates in Vienna attempted to expand its scope of application, both geographically and socially. Late nineteenth-century pitch standardization was congruent with the Euro-American spirit of expansionism—although a nation like Austria was not engaged in the same colonizing enterprises as its Western neighbors. But in increasing the scope of their efforts, the Vienna participants discovered just how difficult it was to control sound across varying geographic and social spaces.

ORDERING THE SONIC WORLD

The Vienna conference's recommendations being nonbinding, it fell to each country to individually pass the measure into law. This process raised important political questions. Although most nations wanted to avoid state intervention, this approach had drawbacks. For example, reminiscent of Hanslick's 1864 article, the German musicians who petitioned Bismarck in 1884 emphasized that while "a number of institutions have introduced the Parisian pitch, the others did not follow." As a result, they declared that they had no other option than to appeal to the Reichskanzler Bismarck. The question then, they continued, was "to determine whether this should be an imperial law or a ministerial decree." While Austria and Germany interpreted the Vienna conference's decision as requiring governmental regulation only, Belgium, Spain, and Italy passed it into law (respectively in 1877, 1879, and 1887).

Determining the legal status of the reform was only the first step toward establishing it as a widely employed practice. As in France, the second question of how to materially embody the standard proved troublesome. During the conference, delegates had examined in detail the matter of manufacturing standardized tuning forks. Following Stefan's recommendations, the delegates declared that forks must have prongs of the same length, set absolutely parallel a centimeter apart. Furthermore, to protect these forks from oxidation and to secure their physical integrity over time, their manufacturers should gold-plate them.[99] Following the French example, Austria, Italy, and Germany all appointed officials for controlling the production of tuning forks in line with the model of the *bureau* established at the Paris Conservatory. In 1887, replicas of Lissajous's office were established at the Physikalisch-Technische Reichsanstalt in Berlin, the physics department of the University La Sapienza in Rome, and the physics faculty at the University of Vienna.[100] Case Western University later became the site of an office dedicated to enforcing standard pitch in the United States, and material evidence also seems to indicate that Moscow created a similar center in 1895: that year, Parisian instrument maker Koenig received a major order from the Russian State University that kept him busy until his death in 1901.[101]

Paris not only offered an institutional precedent for the creation of centers for regulating tuning forks: the French capital also set the standards of acoustic precision in use at these new offices. Koenig provided each of these institutions with the tuning forks required for their reforms.[102] Koenig's business flourished as a result, and the implementation of the international standard in turn owed much to his authority and the dissemination of his products. As early as 1885, the University in Vienna had purchased a set of standard tuning forks from the Paris maker as well as Koenig's masterpiece of acoustic precision, the clock fork. Developed during the summer of 1879 in response to Ellis's allegations that Koenig's diapason normal was inaccurate, the clock fork determined the frequency of a sound by reference to a standard clock. In this apparatus, a vibrating tuning fork drives the clock in the same way as a pendulum. By keeping track of the gains and loss of time between the clock fork and a standard clock, one could determine the frequency of the fork with great accuracy.[103] Over the following decades, Koenig's clock fork became the masterpiece of sound metrology, with the main European centers for the control of tuning forks adopting the device, including in Vienna, Rome, and Berlin, and in cities in Russia and Canada as well.[104] As part of the preparation for the opening of Vienna's new tuning center, the Staatliche Kommission zur Beglaubigung von Stimmgabeln, in 1889–1890, Stefan purchased two additional standard forks from Koenig.[105] Similarly, in creating the Ufficio centrale italiano per il corista uniforme in Rome, Pietro Blaserna purchased a set of Koenig tuning forks (see fig. 2.2) and, a few years later, acquired Rudolph Koenig's clock fork. The physicist Leopold Löwenherz, who oversaw the production of tuning forks at the Physikalisch-Technische Reichsanstalt in Berlin, also recommended the use of these Parisian forks.[106]

By selling his products throughout Europe and North America, Koenig not only exported artifacts but, given that these were loaded with implicit values and techniques, contributed to the dissemination of specific scientific practices on an international scale.[107] From Paris to Vienna, Berlin, Rome, Cleveland, Toronto, and Moscow, the circulation of Koenig's instruments shaped a new culture of acoustical precision. In addition to selling instruments, Koenig's shared acoustic knowledge informed the activities of scientists working at the centers for the verification of tuning forks. For instance, in the years following the opening of the *ufficio* in Rome, Blaserna's assistant conducted a series of experiments on the impact of temperature on tuning

FIGURE 2.2. Koenig's forks used at the Ufficio centrale italiano per il corista uniforme. Museo di Fisica, University La Sapienza, Rome / Cliché Giovanni Organtini.

forks that invoked Koenig's key publications on the topic.[108] Indeed, Koenig himself cultivated relations with several members of these centers.[109]

Despite the intensity of material and intellectual exchanges in the aftermath of the Vienna conference, the participating nations fell far short of introducing the French pitch throughout their territories. Just as in France, the regulation of sound and music revealed the limits of each state's authority: beyond the walls of physics laboratories, the standard adopted in 1885 remained largely unenforced. Having kept detailed records, the Vienna office offers a good starting point for evaluating the impact of state efforts to control sound. The exceptionally well-preserved archives of the institution attest to the care with which the Austrian government planned the implementation of the diapason normal, as well as its failure to impose the measure throughout its empire.

The organization of the Vienna center itself was a testimony to Austria's commitment to the French pitch. Unlike offices in Paris, Berlin, and Rome, where a single physicist oversaw the control of tuning forks with the help of a technician, in Vienna a commission of five members, possessing both scientific and musical expertise, directed the office's affairs.[110] This commission held its first meeting on 9 February 1891. Reflecting its awareness of the practicalities of musical performance, the commission established a double standard of precision. For "precision forks" used in laboratories, the margin of error was set at only "0.2 single vibrations more or less compared to the value of the normal tone," meaning that the pitch of these instruments should be between 869.8 and 870.2 vibrations. For forks intended for musical use, by contrast, the margin was of "one single vibration (½ complete vibration) more or less than the value of the normal tone"; that is, the commission had to tune devices "between 869 and 871."[111] In other words, the commission recognized the necessity of adjusting standards according to their various applications.

Although intended to facilitate access to the new standard, the establishment of the Vienna office created much confusion, as many correspondents of the office believed that it would supply the country with new tuning forks when, in fact, it was only intended to adjust existing devices. In January 1891, for example, Fritz Wagner, a piano tuner from Graz, asked the University of Vienna to send him a fork.[112] Similarly, on 30 December 1892, the mayor of Gutenstein requested two tuning forks for the regulation of "musical exercise and performances, as well as within ecclesiastical circumstances."[113]

In addition to the confusion over the office's function, and illustrative of the considerable preparation involved, it took four years for the *Kommission* to start its regulatory work. On 20 November 1890, the Viennese instrument maker Franz Steflitschek reported that his customers repeatedly asked when they could buy standard tuning forks. Similarly, in February 1891, the workshop Houdek & Hervert in Prague wrote to the office to inquire "when the tuning forks could be submitted" and what the fee would be so that they could fix the price of the tuning forks they would sell.[114] By November, as the production of standardized devices had still not started, the Prague instrument maker Alois Kreidl reported that he was overwhelmed with demands from customers eager to buy standard tuning forks.[115]

Reflecting the *Kommission*'s centralizing ambitions, records for every single tuning fork calibrated by the office were held in the archives of the University of Vienna. As this abundant material reveals, the office tuned more than 20,000 forks over the course of its first twenty years of existence. Despite these efforts, it failed to secure the adoption of a unified pitch throughout the Austrian Empire. In the report of a meeting held in 1912, the government recognized that musical practice was still far from harmonious within the country: even at the Viennese Philharmonic Orchestra, the institution most closely connected to the *Kommission*, the standard was not enforced.[116] Austria's musical reform had delivered a sobering lesson on the limits of both the state and science to regulate musical practice.

In the following decades, several makers and musicians reflected on the failure of the international pitch, not just in Austria, but across Europe. In the early 1930s, one of the members of the Vienna *Kommission*, harmonium maker Teofil Kotykiewicz, wrote a report on the then-current state of pitch unification as well as the many complaints of instrument makers and musicians.[117] To glean the extent of the problem, Kotykiewicz sent a circular letter to "many instrument makers, makers of physical-acoustical instruments, as well as the most significant musical associations," on 9 December 1926. The answers to this inquiry indicated a general commitment to *A* 870, but provided evidence that the measure was often poorly enforced. For instance, the Viennese Schubertbund declared that they always used *A* 870, but the Viennese Männergesangverein conceded that their tuning depended on that of pianos tuned to *A* 876.

Attempted explanations for this acoustic chaos were varied. While some accused solo violinists and wind instrument makers of indiscipline and selfishness, others, such as Alexander Wunderer, identified more fundamental aesthetic problems. An oboist and manager of the Viennese Opera orchestra, Wunderer published a paper entitled "On the Question of Normal Pitch," in which he claimed that pitch standardization should not be addressed from the point of view of equally tempered instruments only, as it had been hitherto:

> The reason for this tendency lies in the peculiarity of the characteristic of orchestral instruments. In contrast to keyboard instruments, the orchestra is not a "tempered" body of sound, but, according to its arrangement, a "purely" tuned one. In practice, this means that each instrument takes the chord tone it has to play in such a way that it blends in perfectly with the harmony. How this has to be done is entirely in the perception of the instrumentalist and can be achieved neither through physical-scientific means, nor with acoustic measuring devices, but only through a lively musical feeling, which must be an unconditional requirement for every musician.[118]

"To be able to produce this absolute purity," Wunderer continued, "every orchestral instrument must be somewhat 'unstable' in pitch, that is, each tone must have the ability to be flexible."[119] Wunderer recognized the utility of a standard and was in favor of the maintenance of *A* 870, arguing that the use of a higher pitch risked "Haydn's string quartets [being] played like a jazz band." Wunderer's comments reveal how, at the turn of the century, pitch standardization began to crystallize anxieties over a changing musical world, characterized by the commercial success of popular music. Yet Wunderer also recommended that the definition of the standard be defined with less mathematical precision, asserting that "what we need is a middle line which is as close as possible to the normal *A*." Just as the Viennese commission for the verification of tuning forks had initially acknowledged in maintaining both a scientific and a musical measure, Wunderer endorsed an expanded definition of the standard. As standardizers looked closer at the implementation of pitch reform beyond the walls of physics laboratories, the limits to regulating music became increasingly apparent.

If pitch uniformity was difficult to achieve in Europe, even in theaters and concert halls, the very locations for which it had specifically been intended, to carry out a reform beyond the "railway network" that Hanslick had spoken of in his 1864 article was even harder. Meanwhile, imperialism was leading to a broader dissemination of the standards adopted by Europe and the United States.

IMPERIALISM AND THE LIMITS OF WESTERN SONIC AUTHORITY

Europe was in no way the birthplace of the concept of tonal uniformity. Pitch standards and technologies embodying them were observed in China long before their emergence in Europe. In first-century China, the emperor Wang Mang used sound as the basis for the standardization of weights and measures. As an interface between heaven and earth, and a central tool for the prediction of seasonal change and astronomical events, sound carried immense sociopolitical implications. Surviving ancient bells embody this knowledge of sound entwined with sociopolitical and cosmological notions of harmony and prediction.[120] Like equal temperament, which also developed in different parts of the world around the same time, there is no evidence of mutual influence. European scholars were aware of Chinese theories and practices, but these did not seem to have an impact on Western conversations about pitch standardization.[121]

In the last decades of the nineteenth century, pitch standards, along with musical languages, repertoires, and instruments, were colonizing tools that transformed global musical practice in ways that are still audible today. In the absence of central authorities charged with regulating musical and sonic practice, however, attempts to unify pitch in the colonies resulted in even less integration than in Europe. In territories only loosely connected to imperial centers like Paris and London, standards were fixed in reference to competing models. Adding to this cultural diversity was the contrast between European and colonial environmental conditions. In short: Europe's colonial and imperial contexts intensified the sociopolitical and epistemological problems that reformers had started to identify at the turn of the century.

Colonial tuning practices, like European ones, were the products of competing priorities. The result was the creation of multilayered musical soundscapes. In places characterized by extreme social and cultural diversity, tuning practices were often more varied than in Western contexts. For instance, the city of Melbourne not only maintained Britain's "Old Philharmonic Pitch" (*A* 452.5) long after the main musical societies and instrument builders of England had adopted the "New Philharmonic Pitch" (*A* 439),[122] but in 1909 the vocalist Nellie Melba introduced the diapason normal at the local orchestra by donating a set of instruments tuned to *A* 435. Like other touring vocalists of her time, Melba preferred to sing at this pitch and made sure the pianos accompanying her were tuned to it. Beyond these practical considerations, Melba's gift to the Melbourne orchestra was intended to establish the city as part of the world's premier musical networks. For all its good intentions, Melba's gift actually diversified the local variety of pitches.[123]

Imperialism also shaped Western pitch negotiations from the outside in. The United Kingdom's refusal to align with Continental tuning practices until 1939 owed much to its imperial commitment and the commercial and cultural weight of its empire. Although the British did not take part in the Vienna conference, the event attracted immense attention among the musical circles of the country—as had the original promulgation of the French pitch in 1859. In 1885, the Royal Academy of Music held a meeting of no fewer than 400 participants from Britain's musical professions to discuss the possibility of adopting the French pitch. The result this time was a consensus in *favor* of the diapason normal, a trend confirmed the following year by a Society of Arts inquiry.[124]

And so, the diapason normal secured considerable support among Britain's musical audiences. However, it failed to become its national measure because of commercial and political divergences over the role of Britain's colonies in the process of pitch unification. During the 1886 Society of Arts' inquiry, composer John Stainer expressed his opposition to the diapason normal, arguing that Britain, given the extent of its empire, should lead rather than follow other European nations, imposing a standard of its own on a global scale.[125]

Despite the Society of Arts' recommendation of *A* 435, the maker of brass instruments Augustus Charles Köhler followed Stainer's reasoning and launched a campaign against the diapason normal, promoting instead the adoption of the "high pitch" in use in the British Army. Known as "Kneller Hall Pitch" after the name of the building occupied by the Royal Military School of Music, this measure was "established throughout the whole universe where English is spoken and British interests predominate."[126] England's main brass instrument makers joined Köhler's initiative and together successfully convinced the government to protect their interests by preventing the introduction of the French pitch in Britain.

In 1895, amid growing controversy about pitch unification, Queen Victoria requested that the diapason normal be employed at the Philharmonic Society and, over the following years, Britain's premier instrument manufacturers and concert societies emulated this royal decision. However, instead of the French pitch, these institutions actually adopted *A* 439 as a standard. In doing so, they followed the recommendation of the piano maker Alfred Hipkins. Drawing on a Society of Arts erroneous translation of the French 1859 decree, Hipkins argued that to produce the sound that France had adopted in 1859 within the conditions characteristic of English concert halls, one needed to adopt a different figure that took into consideration the impact of temperature. His suggested figure rested on two factors: the impact of heat on the pitch of wind instruments, and the choice of 20°C as "the average concert-room" temperature.[127] This proposition drew on a mistranslation of the French decree and further shaped pitch negotiations in interwar America, ultimately driving the international adoption of *A* 440 as a standard.[128]

Wind instrument makers, however, rejected the reform, arguing that "to alter all [army bands'] instruments, or to substitute others for them, would be an enormous undertaking, and there is no indication of the War Office or the officers being prepared to pay for a change of pitch in the Army, to say nothing about the owners of thousands of instruments in wind bands generally."[129] One factor in their resistance was their belief that a lowering of their instruments' pitch would reduce the quality of their sound, their "brilliance." By the turn of the century, Britain's pitch standardization came down to a confrontation between the internationalist ambitions of "civilian" musicians, and the nationalist, often imperialist, conceptions of musical uniformity promoted by wind instrument makers.

These commercial and political tensions resulted in a virtual double standard coming into place throughout the British Empire, one that forced musicians playing in both army bands and orchestras to own two instruments. As the extract from wind-instrument maker Hawkes's catalogs presented in

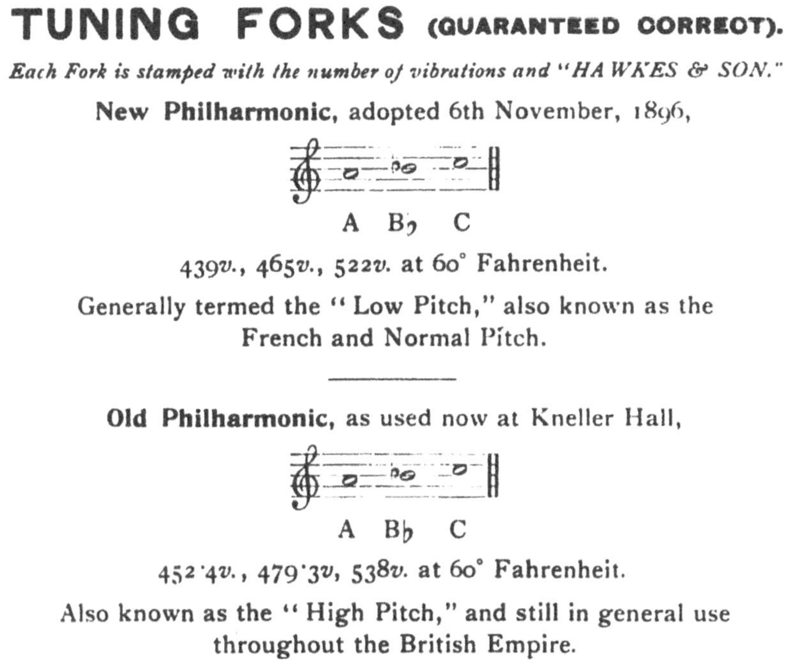

TUNING FORKS (QUARANTEED CORRECT).

Each Fork is stamped with the number of vibrations and "HAWKES & SON."

New Philharmonic, adopted 6th November, 1896,

A B♭ C

439*v.*, 465*v.*, 522*v.* at 60° Fahrenheit.

Generally termed the "Low Pitch," also known as the French and Normal Pitch.

Old Philharmonic, as used now at Kneller Hall,

A B♭ C

452·4*v.*, 479·3*v*, 538*v.* at 60° Fahrenheit.

Also known as the "High Pitch," and still in general use throughout the British Empire.

FIGURE 2.3. Illustrated price list of the Hawkes military band instruments (London: Hawkes & Son, 1912).

figure 2.3 illustrates, it also meant that wind instrument makers had to build three types of instruments—one tuned to the diapason normal (*A* 435), one to its British translation (*A* 439), and one to the old "Philharmonic pitch" (*A* 452.5). This unhappy state of affairs lasted until 1927, when King George V imposed the use of *A* 439 on the British Army, a decision that was not completely enforced until the late 1960s.[130] In other words, British imperialist considerations had shaped the actual pitch not only in the colonies but back in England as well.[131]

By the time that international pitch negotiations began in the 1850s, the dissemination of European musical technology had reverberated across the world, and Western instruments were present on a global scale. Since the sixteenth century, colonial churches had been homes to organs, which kept shaping global soundscapes during the early modern and modern eras. For example, the French writer Gustave Flaubert reported hearing the sounds of an organ accompanying the singing of mass during his voyage to Egypt in 1849.[132] Monumental pipe organs by Europe's eminent makers, such as Germany's Walker, France's Cavaillé-Coll, or Britain's Willis, adorned cathedrals and churches in Asia, Central and South America, Africa, and Australia,[133] while more modest reed organs accompanied the chant of missionaries and lesser acts of worship. Itself the product of a European appropriation of Chinese free reed technologies,[134] the harmonium traveled with Western missionaries to China, Japan, India, the Middle East, and Africa.[135] In addition to these instruments, pianos disseminated Western sounds throughout the second half of the nineteenth century. Just as in Europe and North America, they were central pieces of furniture in colonial homes, where they served as markers of social prestige and delineated gender and racial hierarchies.[136] They were also the cornerstones of musical education in the myriad institutions modeled on Western conservatories that flourished in the late nineteenth and early twentieth centuries. Last but not least, military bands were among the primary sources of global sonic exchanges. In 1846, for instance, only thirty-five of Britain's 112 regiments of infantry were stationed in the United Kingdom, with the remaining seventy-eight deployed throughout its overseas territories, especially India. The musicians not only played a central role in the empire's musical life as a whole, taking part in musical performances held in kiosks, ballrooms, and concert halls, they also left their mark on the musical cultures of the colonies in which they performed.[137]

Musical instruments were thus colonizing tools in the hands of Western missionaries, armies, and businessmen. They were, however, fragile embodiments of empire. First, they were extremely vulnerable to local environmental conditions. Keyboard instruments, for example, although very popular among colonial and local elites in the East, did not maintain their integrity while traveling across regions marked by extreme climatic conditions. Trade literature at the end of the nineteenth century dedicated to the development of overseas business increasingly acknowledged that building pianos and harmoniums for the colonies required using specific materials and techniques suited to colonial climates and tastes.[138] In 1924, for instance, the American trade commissioner in China recommended the exclusive use of "nickeled tuning pins," adding that strings must be "gilded for protection against rust, which prejudices the tone."[139]

Despite the growing care put into the manufacture of instruments designed for tropical climates, pitch in the "extreme" regions remained very much in flux. Although the tone of pianos was relatively stable, that of other instruments was much harder to control. Wind instruments exhibited particular sonic indiscipline. The activities of David J. Blaikley, acoustician and factory manager at Boosey and Co., one of the main providers of musical instruments for the British Army, were typical of Western attempts to rationalize musical practice to fit the variety of environmental conditions around the globe.

A supporter of pitch unification and one of the most trenchant advocates of the diapason normal for commercial reasons, Blaikley conducted substantial experiments on the pitch of wind instruments. The manufacturer was particularly eager to address the challenges of temperature's impact on the pitch of various wind instruments in colonial settings. His extensive experiments showed that the impact varied in proportion to the size of the instrument: whereas environmental variations had a limited impact on the pitch of smaller winds such as the flute, oboe, and clarinet, they had a much greater influence on the larger euphonion and bombardon [see fig. 2.4]. In other words, because "instruments vary in different degrees with temperature, it is impossible that all the instruments in a band can rise and fall exactly together."[140]

What is more, as Blaikley explained, "while the manufacturer may be expected to make [instruments] so that they stand well together at a medium temperature . . . the adjustment necessary to ensure a good ensemble at extreme temperatures must be left to the judgement and experience of the players."[141] Blaikley's careful observations once again revealed the limits of Western ambitions to globally regulate sound frequencies. The army may have been the ideal institution through which to implement pitch reforms, but the often-extreme climates encountered in colonies profoundly inhibited the Empire's power to deliver a uniform sound.[142]

Pitch reform was part and parcel of the imposition of imperial standards on non-imperial cultures,[143] but European pitches hardly ruled these territories, even if some global sound studies have promoted the idea that "by the turn of the twentieth century, march-based instrumental conventions had defined the international sound of the West."[144] Because of climatic challenges, I would in fact argue that nowhere were the limits of standardizers' ability to control sound more audible than in colonial contexts.

Hybridization and resistance further contributed to the alteration of Western pitches along the course of their travels to the colonies. The global history of concert pitch is best understood in the context of the broader debates that started to surround the use of Western scales and instruments at the turn of the nineteenth century. With the emergence of nationalist movements rejecting colonial occupation in various parts of the East, the introduction of Western instruments and tuning systems became the subject of heated debates, although concert pitch does not seem to have been an object of the controversy. Meanwhile, indigenous performers' appropriation of Western

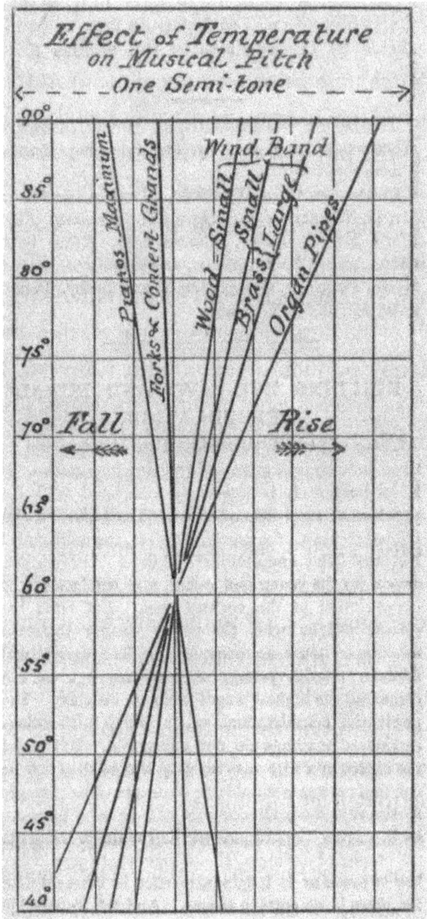

FIGURE 2.4. "Cantor Lectures. Musical Wind Instruments. By D. J. Blaikley. Lecture IV. Delivered December 19, 1904." *Journal of the Society of Arts* 53, no. 2721 (1905): 181. Courtesy of JSTOR.

standards produced hybrid sounds. In India, for example, where the harmonium became an integral part of Hindustani musical practice, the instrument became the focal point of heated debates between British and Indian musicologists. Scholars eager to distinguish Western music from its Indian counterpart produced a double ontology: "the West" was construed as synonymous with discrete units and equal temperament, whereas Indian music was portrayed as defined by continuity and a twenty-two-*sruti* intonation system.[145] As the main embodiment of this alleged Western musical system in India, the harmonium was deemed incompatible with the promotion of national canons. The Indian poet and composer Rabindranath Tagore considered the harmonium a tool of British domination and rejected it as unsuitable for Indian music. His imagination of "a free modern India," that is, one that would develop "an indigenous modernity, separate from Europe," included liberation from Western temperaments.[146] As he boasted: "I practiced my songs with my tambura resting on my shoulder; I did not subject myself to the slavery of the keyboard." Europeans, on the other hand, Tagore continued, with "the hideous structures where their children are interned when they take their lessons" and their "square houses with flat, straight wall surfaces, pierced with parallel lines of windows . . . are caged in their lifetime." These buildings "are not modern, but merely European," he concluded. "True modernism is freedom of mind, not slavery of taste."[147]

Between 1940 and 1971 the harmonium became subject to a ban from All India Radio, the country's main broadcasting station.[148] As these scholarly debates proceeded, harmonium players developed strategies to overcome the divisions of the scale in tones and semitones, including ornamentation, variations in air pressure to alter pitch, and the manufacture of "special reed-banks tuned for specific ragas and specific keys."[149] Whether harmoniums and *sruti* boxes should utilize an equal-tempered scale and concert pitch *A* 440 is a question still debated today.[150]

The introduction of Western keyboard instruments created similar controversies in the Middle East. The introduction of the harmonium and piano in that part of the world resulted in heterogeneous combinations of traditional Arab and European instruments and musical systems, with the voice, *ʿūd*, and violin articulating the intervals of the *maqāmāt*, while the keyboard played in equal temperament.[151] Whether this was a good thing for Arabic music became the crux of heated debates during a 1932 Congress in Cairo, where a collection of scholars from Europe and the Middle East discussed the past, present, and future of "Arabic" or "Oriental music."[152] Questions of a unified scale for Arab music and the role of Western musical instruments, especially the piano, stirred up considerable controversy.[153] The roles of proponents and opponents of Westernization were reversed: European comparative musicologists rejected the use of the piano in Arab music, whereas Middle Eastern scholars considered it a powerful tool for standardizing intonation and helping establish an Arab musical scale.[154] The question of concert pitch per se did not arise in these various debates, but dissemination and alteration of Western tuning systems had a material effect on pitch standards. The cultural context reshaped Western sounds. The result of the spread of Western musical instruments in the colonies thus was not sonic uniformization, but rather an increased diversity of musical modes and pitches.

During the age of comparison that emerged from the acceleration of global exchanges, pitch not only became a measure of environmental differences between diverse regions of the world—it also served as a yardstick to assess socio-cultural differences. Once again, Ellis's work offers a good example of this phenomenon. In 1885 Ellis, pursuing his investigations into musical pitch's historical variations across Europe and the United States, published an article entitled "On the Musical Scales of Various Nations." In it he explored the diversity of musical systems from North Africa to India, Japan, and China.[155] While most nineteenth-century music scholars considered the structure of the diatonic scale as universal, Ellis, drawing on his measurements of the pitch of non-Western musical instruments, adopted a relativist stance, insisting that scales were "just various."[156]

Indeed, scholars of pitch were increasingly recognizing standards of pitch as the audible signs of cultural differences between peoples. In 1894, the American archaeologist Alice Cunningham Fletcher identified "the lack of definite pitch" as "the most striking peculiarity" of Native American music:

> They have no mechanical device by which to establish or to promulgate such a pitch. Where a standard pitch exists and its use is enforced, social conditions are implied that do not obtain in an Indian tribe. The Indian starts his song where the natural quality and his present mood renders it easiest for him to sing it. A tenor will naturally sing upon a higher pitch than a bass; a soprano will differ from a contralto. The pitch of a song depends upon the individual; but it in no way affects the intervals of the tunes.[157]

In this way, a stable pitch "or the absence thereof . . . became a measure for comparing different levels of evolution, historical development and positionality on a global scale."[158] The institutionalization of comparative musicology in the opening decades of the twentieth century cemented this new boundary between musics of the world. The most iconic products of German *vergleichende Musikwissenschaft*, the recordings produced by the Phonogramm-Archiv from 1900, reveal just how central pitch became to global comparisons of sound. They systematically start with a mouthpipe giving an *A* and provide the listener with a point of reference to position the sounds of "others" in relation to the *Normalton A* 435.[159]

As awareness grew of the relativity of Western musical sounds and the historic and geographic contingency of attempts at standardization, pitch, like tonality and the tempered scale, became reified as a natural feature of European art.[160] Nonetheless, at the beginning of the twentieth century the sonic unification of Europe was far from resolved. And in the places where the diapason normal *did* regulate musical practice, the transformation of sonic practices had been the product of considerable labor.

CONCLUSION

The pitch standardization of Europe was a slow, largely incomplete process. In the nations where it was introduced it secured financial support, skills, and infrastructures and reshaped the scientific institutions and practices of these countries, stimulating the circulation of acoustical instruments, knowledge, and procedures on an international scale. Given the ubiquity of tuning forks in late nineteenth-century experimental culture, pitch reform had implications that transcended the realm of musical acoustics.

The travels of the French pitch reveal the contradictions between universalism and localism in musical practice. As the idea of a standard pitch spread across new territories, it lost both its authority and integrity. The circulation of the French pitch across Europe and its colonies revealed that, at a fundamental level, there could be no such thing as an absolute point of reference in the realm of sound. Far from establishing a universal measure, the global dissemination of the French pitch demonstrated sound's social and climatic relativity.

CHAPTER THREE

Retuning the World

Transatlanticism and the Defeat of the French Pitch

> It's a complex thing, being an American, and one of the responsibilities it entails is fighting against a superstitious valuation of Europe.
>
> HENRY JAMES

How did concert pitch go from *A* 435 to *A* 440? And what does it have to do with temperature and popular music? In tracing the transatlantic story of the French pitch, this chapter highlights the American roots of *A* 440 and shows how the world was retuned to the new standards of the United States' emerging entertainment industry.

In *The Soundscape of Modernity*, historian Emily Thompson situates the birth of "modern acoustics" in the 1900 opening of Symphony Hall in Boston.[1] A few decades earlier, the predecessor of that venue was the site of another important event in the history of sound in the United States: on 2 November 1863, a great organ built by the world-leading German maker Eberhard Friedrich Walcker was inaugurated with great pomp in the Boston Music Hall (see fig. 3.1),[2] "in the presence of an audience, which, for distinction, beauty, and fashion, has seldom if ever been matched in our country." The most illustrious American actress of the century, Charlotte Cushman, opened the ceremony with the declamation of an *Ode* recounting the struggles of both Bostonians and Americans generally during the Civil War and celebrating the future "Nation's Victory."[3]

Although organ building was a dynamic field in the United States, and specifically in Boston,[4] the instrument had been commissioned from a German builder in Ludwigsburg and transported from a Rotterdam factory aboard

FIGURE 3.1. Great Organ, in Music Hall, Boston, MA. Gift of Weston J. and Mary M. Naef. Digital image courtesy of the Getty's Open Content Program.

the steamship *Presto*. Presented as a symbol of American grandeur, the organ might easily have been interpreted as a "monument to America's cultural inferiority,"[5] especially since the musical program of the evening offered a celebration of European musical genius. Following the Ode's reading, a recital highlighted the various elements of the instrument through a selection of pieces epitomizing the "earnestness" of the organ's commissioners and their ambitions for the city's musical scene, and with it, Boston's moral, social, political, and religious life. Under the attentive eyes and ears of Bach, whose bust was placed on top of the organ, and Beethoven, whose statue had been installed on the stage in 1856,[6] organists from Boston and New York explored the instrument's potential through a program centered on historic German composers. The production was not only one of the earliest examples of an entirely instrumental organ concert, but one of the first performances of Bach's works in America.[7]

Beyond the sounds of great German masterpieces from the past, the organ was in fact tuned to the tone of the new, yet simultaneously "old"[8] French standard, adopted just a few years earlier: Boston's new organ was tuned to *A* 435. Although originally intended to match the higher pitches of Boston's orchestras when the contract was first signed in 1857, the tuning was changed immediately after the announcement of the French diapason normal in 1859.[9] The adoption of the new French standard, however, would be short lived. Despite this setback, the French pitch exerted increasing influence in the United States over the following decades.

THE SOUND OF EUROPE: THE FRENCH PITCH ARRIVES IN AMERICA

The rise and fall of the Boston Music Hall's organ illuminates the conditions that first favored, then challenged, the diffusion of the French pitch in the United States. Although the installation of the instrument was initially perceived as the beginning of a new era in the history of American organ building, less than twenty years later the masterpiece was dismantled, sold, removed from the hall, and placed in storage. Yet despite the organ's fall from grace, the history of Boston's great organ and its entanglement with the import of European tuning practices reveals the cultural and social resonances of the French pitch as it arrived in the United States. The French norm fit within broader American ideas of "Europeanness" associated with the celebration of classical music and drawing on its religious, moral, and political value. Local actors' attempts to enforce the norm in Boston were part of a wider effort to elevate their city's cultural status.[10]

In the mid-nineteenth century, Boston had become a center for the cultivation of classical music. Following the 1815 creation of the Handel and Haydn Society, which aimed at cultivating "the love of good music and a better performance of it,"[11] the Academy of Music provided Bostonians with numerous performances of Beethoven's symphonies, especially his Fifth.[12] The Musical Fund Society Orchestra also offered Viennese orchestral music to large audiences at Tremont Temple; and the Germania Orchestra, a popular touring group of twenty-four musicians from Berlin, settled down in Boston in 1851.[13] In addition, in 1837 several graduates from Harvard University and members of the Pierian Sodality created the Harvard Musical Association, a group with the goal of raising the standard of musical taste at Harvard and creating a music department—which was indeed founded thirty years later.[14] The Harvard Musical Association set up a library and organized chamber music concerts. In 1852, one of the association's founders, John Sullivan Dwight, established the *Journal of Music.* A Unitarian minister, Dwight intended to promote his transcendentalist conceptions of music and utopian views of society through musical criticism. The *Journal* played an important role in shaping the new modes of listening that were articulated through the rise of "serious" music on American stages.[15]

The cultivation of this new taste for classical music took shape as well through the project of building a hall specifically for musical performances that was equipped with a great organ. Jabez Baxter Upham was the driving force behind this project, initially putting the idea to the Musical Fund Association in 1850 and then leading the Acoustic Committee appointed by the Harvard Musical Association. A physician specializing in the study of typhus fever,[16] and a tireless community organizer, Upham played a crucial role in the orchestration of Boston's musical life. He was the president of the Boston Music Hall Association, dedicated to constructing and then managing the venue from its creation in 1850 until 1880, a function he combined with the presidency of various local musical committees.[17]

It took the Harvard Musical Association only two months to raise the $100,000 required for the construction of Boston Music Hall, a testament to Boston socialites' commitment to the construction of the new music venue. The building reflected Upham's desire to establish classical music performance on a sound scientific basis. Built by George Snell (1820–1893), the hall was a rectangular room that could accommodate an audience of 2,585. With a high ceiling fully forty feet above the floor of the upper balcony, the hall's architecture followed Upham's teachings on sound propagation, and drew on conclusions by the investigation of the various parliamentary committees charged with the reconstruction of Britain's House of Commons at Westminster, during the 1830s and 1840s.[18] Upham declared that music needed "a fitting abode,"[19] that its effects depended on "the architectural qualities of the building."[20] He argued that

> There is no sublimer sound than the mingling of a thousand voices and instruments in an open field . . . . In the construction of a Concert Room, therefore, our efforts should be directed to overcome, as far as possible, the imperfections to which every musical performance is necessarily subjected when confined within the four walls of a building. Theoretically, could we

> secure the ready passage and equal diffusion of sound over the whole apartment, without the intervention of reverberation or disturbing echoes, we should have a *perfect* Music Room, in every part of which the auditor would hear with equal distinctness and accuracy.[21]

Upham's insights on architecture and music were shaped by his religious, social, and aesthetic views. His model for the "*perfect* Music Room" was Westminster Abbey, a building whose main merit from a sonic point of view was that it had hosted the imposing commemorations of Handel in 1784. Although not primarily designed for musical performances, Upham alleged that "the ancient cathedrals are most benefiting the majestic movement of the oratorio."[22] Upham hoped to revive the oratorio in the United States, and he argued that American halls should be built to replicate the forms of the European churches where the genre had emerged and blossomed. However little George Snell's building resembled Westminster Abbey, it fully embraced the idea that music performance structures should have vast proportions akin to great European ecclesiastical architecture (see fig. 3.2).[23] Upham's observations appeared increasingly relevant given the introduction of an organ a few years later: for what could evoke European cathedrals better than this fundamental instrument of Christian worship?

Upham's views on acoustics were inscribed with religious and sociopolitical values, but also shaped by his expertise in the field of infectious diseases. His discourse on the Boston Music Hall especially reflected the rise of emerging concerns about hygiene, which were entwined with the acoustic knowledge of sound propagation. Upham outlined how in "a room containing a crowded auditory, artificially lighted and warmed in the usual manner, the air becomes rapidly loaded with the products of respiration and combustion, and, too often, by the addition of coal gas from the furnace flues." In contrast, he continued, "the system of lighting adopted in the Boston Music Hall is such as to avoid entirely the acoustic disturbances above mentioned while, at the same time, it acts as the effectual motive power to the ventilation arrangements of the Hall."[24] In turn, Upham incorporated new technologies of sound into his practice as a physician. For example, in 1858, in collaboration with an engineer and several medical colleagues, he developed the "sphygmosphone," an apparatus that allowed a physician in Cambridge to examine remotely a patient in Boston.[25] Likewise, Upham wanted the most current technology to advance the development of Boston's musical life. To this end, he gathered information about American builders, then set off on a tour of Europe in 1853. In total, Upham spent five months abroad, reaching

FIGURE 3.2. "View toward the stage (ca. 1856)." Methuen Memorial Music Hall Archives.

Liverpool at the beginning of May and returning to Boston at the end of September. It is not clear whether the primary goal of his trip was a comparative investigation of typhus epidemics in London and Dublin, something Upham accomplished during the first weeks of his stay,[26] or his subsequent tour of Europe's premier organs.[27] Whichever, Upham came home with proposals for the Boston Music Hall organ from two prominent builders: William Hill, maker of the greatly admired Birmingham Town Hall organ, and Walcker, at the time one of the most famous builders in the world.

The trustees of the Music Hall agreed in June 1856 to start raising funds for an organ. Shortly afterward, Upham embarked on a second journey to Europe to secure a deal with an organ builder. Of all the organs he listened to, Walcker's instrument in the cathedral of Ulm impressed him most. "To have examined further," Upham contended, "would have been like returning to the ground after climbing the summit of Mt. Blanc."[28] Nevertheless, he wanted the Boston organ to be a comprehensive representation of the most advanced instrument in the field, and so he asked Walcker to accompany him to Paris and London to see whatever Germany's two rivals had to offer. Like classical music in the United States as a whole, Boston's great organ was something of a "mutant transplant," as historian Joseph Horowitz put it,[29] a French-German synthesis of combined European achievements with an "atypical" technology.[30]

The organ introduced into the Boston Music Hall came complete with another European enhancement: the French pitch. In Boston as in Europe, the diapason normal represented a significant decrease from pitches then in use in the city and across the nation. The Handel and Haydn Society's orchestra played at *A* 449, and most organs in the country exhibited similarly high standards.[31] On 3 January 1864, the organ accompanied the annual performance of *The Messiah*, signaling its main musical function: to highlight the classical oratorio. Unfortunately, according to the report published in *Dwight's Journal of Music*, the musicians "began out of tune, and several times were guilty of an 'uncertain sound,'" which was a "sin they too might charge with reason on the Organ; it being awkward to adapt their instruments at once to its low pitch (the new French pitch) so effectually as to feel at home in it."[32] A worse problem during this concert, however, proved to be synchronization. The instrument was slow to respond to the pressure of the performer's fingers on the keyboard, a problem that would attract the mirth of local organists for several decades.[33] The following year, the Handel and Haydn Society (of which Upham was also president) found a temporary solution to the tuning problem by purchasing a new set of woodwind instruments in collaboration with the Music Hall Association.[34] Despite these changes, *Dwight's Journal* noted later that during an oratorio concert, the orchestra had difficulty playing "in perfect unison with the Organ pitch" and "some of the instruments sounded rough."[35] Musicians who performed as well with ensembles other than the Handel and Haydn Society may have preferred to keep their old instruments rather than accustom themselves to the lower-pitched ones.[36]

Tuning the orchestra to the organ's pitch was a pragmatic response to a practical difficulty, since retuning the organ would have been far more work. But this was also highly ideological,[37] as Upham and Dwight's circle was well aware of ongoing European debates over the tuning question. Since 1860, *Dwight's Journal* had echoed overseas conversations around the standardization of pitch. Reports on the work of the British Royal Society of Arts' committee of 1859–1860 appeared regularly in *Dwight's Journal*, as did accounts of the various steps taken by German, Russian, and Belgian cities and institutions.[38] The journal also reviewed studies on musical pitch, including a call for information about the status of the standard throughout the United States made by a Brooklyn professor of music in 1865.[39] In 1863, before the inauguration of the organ, Dwight's introductory address to the musical season began with a few hopeful lines about the impact of the instrument's pitch on the performance of classical masterpieces, especially those with high notes that made them challenging for singers, particularly Beethoven's Ninth Symphony.[40]

Like the hall and the organ, the French pitch was part of the "reform" promoted by Dwight, Upham, and their associates. Some years later, the organist and composer John Knowles Paine (1839–1906), who held the first teaching position in music at Harvard in 1875 and gave his name to the university's music hall in 1914, summed up these views in a lecture reprinted in several newspapers. Paine employed common arguments in favor of the low pitch, identifying the "gradual raising of the musical pitch, which has been growing higher and higher ever since modern instrumental music gained such prominence," as a "cause of the decline of the vocal art."[41] To protect the monuments of music history, whose names would later decorate Paine Hall's ceilings, it was essential to adopt the French pitch.[42] Paine's call for the enforcement of the overseas standard in Boston came after a failed but very active campaign, launched by Dwight in April 1869, to "bring [Boston] in line with the movement."[43] These efforts were almost certainly triggered by the intensification of debates in Britain, and further reports in *Dwight's Journal.*[44] To raise money to buy yet another set of wood instruments tuned to *A* 435, the committees of the Harvard Musical Association, the Handel and Haydn Society, and the Boston Music Hall Association organized a fundraising event on 20 May 1869: a concert entitled "Normal Diapason!" (fig. 3.3).

Besides raising money, this concert was intended to convey a concrete sense of the question of pitch by devoting specific attention to retuning during the performance. Philanthropic musical events sometimes gave a musical representation of the natural disasters for which they were held; to encourage donations for victims of floods or storms, organists would perform improvisations imitating the sounds of natural catastrophes. In a similar way, the "Normal Diapason!" concert offered the audience a spectacular encounter with pitch. Though otherwise conservative, the program had a surprise in store for the listeners. *Dwight's Journal* reported that, for the performance of Mendelssohn's hymn, before the organ and the voices came in,

> the instruments had to be tuned down (by such imperfect means as were available) to the organ or French pitch; nor was there much attempt to hide

NORMAL DIAPASON!

The Handel and Haydn Society,
Harvard Musical Association, and
Boston Music Hall Association,

THROUGH A JOINT COMMITTEE

Representing each Organization, present to their Patrons the following Programme, in aid of their efforts to establish the

NORMAL DIAPASON, or French Pitch,

for all Orchestral or Choral performances,

—AT THE—

BOSTON MUSIC HALL,

ON THURSDAY AFTERNOON, MAY 20th, 1869.

The Committee take pleasure in announcing that

MISS ADELAIDE PHILLIPPS,

MISS J. E. HOUSTON, AND

MR. WILLIAM J. WINCH,

Have kindly volunteered their valuable aid for the purpose named.

B. J. LANG,........................Organist.

CARL ZERRAHN,.....................Conductor.

TICKETS, (with Secured Seats,) at ONE DOLLAR each.

Doors open at 2½ o'clock. Concert will commence at 3½ precisely.

E. L. BALCH, Printer, 14 Kilby Street.

FIGURE 3.3. "Normal Diapason!" Methuen Memorial Music Hall Archives.

> the awkward process either from sight or hearing. Such a tuning up [*sic*] as there was! prolonged and mystifying; one might have fancied some Lisztian or Wagnerian poem of the Future suddenly interpolated. Someone, plainly, from the "rural districts," asked his companion, "What is all this?" "Oh" replied he, "this is where the normal diapason comes in." The little episode was a good practical demonstration of the need of the reform.[45]

With this comment Dwight at once dismissed the less sophisticated segment of the audience and German modern music, and celebrated the concert's pedagogical achievement. Again, however, Dwight's efforts failed to achieve the intended results. As in 1865, instruments purchased with the money raised by this concert "vanished out of sight and knowledge,"[46] and during the following Christmas oratorio performance, "the chronic incongruity of old and new pitch marred the euphony."[47]

The 1869 diapason concert was the last recorded effort by Boston's concert societies to spread the French pitch. By 1871, the standard's defeat was complete; the organ was retuned to match the orchestra's pitch, and not vice versa.[48] This retuning was a prelude to the instrument's material decay. Played less and less throughout the 1870s, it was abandoned altogether following the creation of the Boston Symphony Orchestra, whose repertoire moved away from oratorio in favor of symphonic and, increasingly, modern music. Given the silencing of the organ and the orchestra's move toward contemporary musical works, the very presence of Walcker's masterpiece on the hall's stage became a bone of contention. Musicians were soon demanding the entire space for themselves. In 1884, the organ was sold and put into storage at the New England Conservatory. Finally, the construction of tennis courts necessitated its sale at auction. Bought from Walcker for $60,000 in 1863, the organ fetched only $1,500 in 1897.[49]

TUNING TO EUROPE: PIANO MANUFACTURERS, PITCH DATA, AND THE DIAPASON NORMAL

Pitch was not just a concern for Boston's musical culture: it became a critical issue for the booming musical market in the United States at the turn of the twentieth century. With the rise of professional orchestras in the 1880s, a number of prominent American musical organizations adopted the French pitch in order to facilitate performances by European musicians on tour. In addition, musical unions were starting to emerge in the United States by the time of the 1885 Vienna conference. These often perceived pitch standardization as a mechanism that would encourage the emergence of a national musical market based on European standards. The National League of Musicians, established in 1886, resolved at its first meeting in Chicago in 1887 to promote the French standard.[50] In 1886, the National Music Teachers' Association convened a Committee on Pitch,[51] and called attention "to the fact that last season at Vienna a convocation of the leading musicians" had officially adopted the French pitch, as had "many musical organizations, societies and individuals in England." The committee therefore declared *A* 435 "the only authoritative musical standard of the civilized world." Finally, the new American Federation of Musicians proposed standardizing pitch in its 1896 program, and adopted *A* 435 as its standard in 1902.[52]

The most significant efforts to reform pitch in the United States came from piano and organ makers. These actors had a material motivation for being the principal movers of the idea of standardization to the French pitch. First, being the producers of fixed-pitch instruments, they could expect a lot of benefits from unification. What is more, as these manufacturers began to market and sell their products across Europe and its colonies, they became eager to tune their instruments to the French pitch. During a meeting of the newly created Piano Manufacturers' Association in New York on 31 March 1891,[53] makers had agreed that a common standard would help them export their products throughout the country and abroad and appointed a committee to resolve this question, with William Steinway (1835–1896) as chair.[54] The committee included several of the country's most prominent manufacturers,[55] who believed that rationalization of their production was crucial both to industrialization and for securing new markets in Europe, and recommended the French pitch, which the Piano Manufacturers' Association accordingly adopted as their national standard[56] on 6 November 1891. Given the ubiquity of pianos in the United States at the time and the instruments' power to set the tone with their fixed pitches, the decision was a crucial step toward the introduction of the measure throughout the country.

The Piano Manufacturers' Association's decision to introduce industrial regulations among American instrument makers was the result of intense interactions between industry and science, and informed by the transatlantic spread of acoustical knowledge, musical instruments, and devices of scientific measurement. From 1891 to 1892, Levi K. Fuller (1841–1896), secretary of the association's pitch committee, took the initiative in the inquiry's proceedings. An organ maker from Vermont, passionate astronomer, tuning fork aficionado, and tireless politician, Fuller collaborated actively with the physicist

Charles R. Cross.[57] Together, they transformed the tuning practices of American instrument makers so as to reflect, and be calibrated to, European acoustical and musical procedures. Examining the dialogue between Fuller and Cross provides rich insights into the historical processes of pitch standardization and shows how exchanges regarding aesthetics, science, and industry shaped this second phase of the French pitch's spread in the United States.

Fuller first encountered Cross in 1882, when he asked the physicist, a professor of physics at MIT,[58] to measure tuning forks that he had in his possession. Their subsequent exchanges appear to have commenced in March 1883. In a letter responding to Cross's request for information on the forks sent for examination, Fuller provided him with details about the devices "used by several manufacturers, orchestras, and artists."[59] In reaching out to Cross, Fuller was addressing one of New England's experts on acoustical matters, and someone already fully engaged in conversations about pitch standardization. In 1872, Cross had given public lectures on acoustics and in 1874 started collaborating with Alexander Graham Bell, inviting him to use the acoustics resources of the laboratory.[60] Cross retained close ties with the Bell Telephone Company during the 1880s and 1890s, becoming the company's "chief expert" and serving as a witness in one of the many patent trials faced by Bell.[61] Since 1868, MIT had owned several examples of European acoustic apparatus;[62] when Cross became head of the physics laboratory, he continued to update its acoustical equipment. In 1883, he oversaw the move of the acoustics and electrical laboratory to a new building that provided greater space (see fig. 3.4).

Cross developed an interest in musical pitch over the course of several experiments on the sounds of language and the perception of tone.[63] In 1880, in collaboration with the son of local piano maker Henry J. Miller,[64] he measured the pitch of dozens of tuning devices and instruments employed by piano and organ makers from Boston. Although the exact circumstances of this experiment are unrecorded, it seems to have been directly connected to the introduction of the great organ, in Boston.[65] In contrast to Upham and Dwight's aesthetic, moral, and religious concerns, Cross and Miller were more engaged with the industrial and commercial ramifications of pitch. Cross's attention to pitch was part of MIT's efforts to support US industry, and particularly piano and organ making, through engineering.

To conduct their study, Cross and Miller's measurements made use of European acoustic knowledge and scientific instruments. Koenig, especially, was a central figure in the development of Cross and Miller's study. His publications and his instruments, some of them repurposed from previous exper-

FIGURE 3.4. "The acoustics laboratory at MIT, about 1890." Courtesy MIT Museum.

iments on speech, were fundamental to the Bostonians' investigations.[66] In turn, Cross and Miller's study contributed to the United States' growing influence within the field of acoustics, demonstrating the Americans' skill in handling sonic phenomena and their determination to secure high standards of scientific objectivity.[67] Pursuing the earlier studies of Scheibler, Lissajous, and Ellis, Cross and Miller's article was also the first contribution to the study of tuning practices in the United States.[68]

Using the method of beat counting—recording with the help of a chronometer the number of interferences caused by the superposition of two sounds slightly out of tune—Miller and Cross measured thirty-three tuning forks.[69] A little over half of these actually "represented" standards of musical pitch in Boston, and the rest included measurements of forks associated with Steinway in New York, Theodore Thomas's orchestra in Chicago, and several European cities and institutions.[70]

Cross and Miller's measurements represented a shift in American pitch conversations away from aesthetic, social, and religious concerns, and toward scientific and industrial considerations. Yet they also revealed that the two remained closely connected. To be sure, Cross and Miller's paper fit within MIT's commitment to helping US industry through engineering work, offering suggestions for a rationalization of instrument making. The authors particularly called for the adoption of a single standard, rather than the then contemporary practice of using two, one for an *A* and one for a *C*. In addition, Cross and Miller drew attention to the lack of accuracy in Boston piano makers' practices, emphasizing how much the pitch of their tuning forks varied depending on the ambient conditions of the rooms in which they were used. One should be aware, for example, the authors argued, of the difference between the pitch of a tuning fork employed in a church in the winter versus one in a concert hall. In parallel to these scientific and industrial considerations, however, the paper reflected prominent cultural overtones. In the same fashion as their European counterparts, Cross and Miller were especially keen to draw attention to pitch's variations over time. To this end, they presented their data in an ascending progression so as to demonstrate that in the good city of Boston, pitch was both chaotic and rising. Cross and Miller argued in favor of pitch regulation in the United States.[71] Drawing on an article that Ellis had published in *Nature* just a few months earlier, and in a way reminiscent of the practices of European physicists before them, Cross and Miller engaged with the aesthetic debates triggered by the history of the Music Hall organ, these being familiar to them as Bostonians.[72] They explained how

> It is important . . . that not merely the works of Händel and Haydn, but those of all the founders of modern classical music, were written to a pitch (A_3, 423 vibrations, or thereabouts), considerably lower than the lowest orchestral pitch at present used in any country, lower even than the physical pitch. The present high pitch used in Boston, like that of New York and London, is over a diatonic semitone sharper than this classical pitch.[73]

Cross and Fuller's collaboration brought this mix of industrial and cultural considerations articulated by Cross and Miller onto a whole new level. With his collection of tuning forks, Fuller enabled Cross to pursue his measurements,[74] and thus document American tuning practices. In 1891–1892, the data they created found a direct application when it became part of a publicity campaign aimed at persuading piano and organ builders, as well as the general public, of the urgency for pitch regulation. In Cross, Fuller found an ally who provided him with both the scientific and technical means to objectify American tuning practices, and sufficient authority resulting from such acoustic mastery, to convince makers of the importance of adopting a common pitch.[75]

Following the Piano Manufacturers' Association's appointment of a committee to investigate the issue of pitch, Fuller initiated an active campaign of "agitation" to raise awareness of the value of pitch standardization for American trade and thus secure consensus for a unifying measure. During the spring, summer, and fall of 1891, Fuller distributed "circulars" throughout the nation's musical communities as well as to the press.[76] Intended to "educate" musical audiences on the question of pitch reform, these documents provided American readers with a historical account of earlier European pitch negotiations, as well as scientific and technical knowledge of acoustics.[77] In parallel to the dissemination of these texts, Fuller corresponded with a broad range of actors throughout the musical world, including manufacturers, orchestra directors, heads of local and national musical unions, and singers, in an attempt to convince them of the necessity of adopting the French pitch.[78] A broad consultation, he maintained, was the best way to reduce the opposition to the reform. As he put it in a letter to Steinway: "There is so much in human nature that has to come to the surface whenever any opportunity occurs, that the best way to head [opponents] off is to consult them and get them to agree with you."[79]

The collection and measurement of forks was a crucial part of Fuller's public campaign. Keen to obtain as comprehensive an account of the state of pitch within the United States as possible, Fuller accumulated data throughout the months of the committee's work. In addition to his "Circulars," he wrote to manufacturers requesting them to send him tuning forks representative of the standard to which they tuned their instruments. At the end of September 1891, for instance, he wrote to a maker in San Francisco, asserting that "he was very anxious to have California represented" in his collection.[80] Fuller subsequently presented the measurements from this national survey of instrument builders' forks within a table, published in his tenth "Circular" (fig. 3.5). Below this appeared three other tables, respectively giving the pitch values of forks from Fuller's collection, the experimental data that Cross and Miller had published in 1880, and, finally, "statements of parties without forks," composed of the responses of manufacturers who had failed to attach a tuning device. The statistical evidence displayed mainly represented the pitch used in the country's largest factories, concentrating on the East Coast and, specifically, New York.

Fuller not only mobilized numerical data, but also transformed tuning forks themselves into pieces of evidence. Recalled Fuller: "On Friday [6 November 1891] I spoke at some length on musical pitch before the Piano Mfrs . . . . I purchased in N.Y., Chicago, St. Louis and Detroit tuning forks of

TABLE NO. 1.

Forks received in response to the committee's circular No. 1, and rated by Prof. C. R. Cross:

	C4 Fork.
Roosevelt Organ Works, New York	516.8
J. H. & C. S. Odell, New York	520.6
J. H. Foote, New York	535.4
George Steck & Co., New York	539.5
Wm. E. Wheelock & Co., New York	539.6
Decker Brothers, New York	539.9
Estey Piano Company, New York	540.
R. M. Bent & Co., New York	540.1
P. G. Mehlin & Sons, New York	540.2
Pease Piano Company, New York	540.4
Newby & Evans, New York	541.
Hazleton Brothers, New York	542.3
Decker & Son, New York	542.3
Mason & Hamlin, Boston	537.7
C. C. Briggs & Co., Boston	537.7
Vose & Sons, Boston	537.8
McPhail Piano Company, Boston	538.6
Chickering & Sons, Boston	538.8
Chickering & Sons (Low), Boston	521.2
C. B. Snyder, Winfield, Kan	518.4
Jewett & Co., Leominster, Mass	533.
Gallup & Metzger, Hartford, Conn	534.5
Shoninger O. & P. Company, New Haven, Conn	534.5
Boardman & Gray, Albany, N. Y	539.5
Leicester Piano Company, Leominster, Mass	550.9
Estey Organ Company, Brattleboro, Vt	517.3
Estey Organ Company, Brattleboro, Vt	522.
Estey Organ Company, Brattleboro, Vt	540.
Wm. Knabe & Co., Baltimore, Md	508.8
Wm. Knabe & Co., Baltimore, Md	541.3
Wilcox & White Organ Company, W. Meriden, Conn	532.6
Lester Piano Company, Philadelphia, Pa	547.8
Clough & Warren Organ Company, Detroit, Mich	538.6
Atlanta Piano Company, Atlanta, Ga	539.1
A. B. Chase Piano Company, Norwalk, Ohio	540.8
Sherman, Clay & Co., San Francisco, Cal	545.4
J. & C. Fischer, New York	526.2
Francis Bacon, New York	536.4
Albert Weber, New York	537.5
Behning & Sons, New York	538.1
Behr Brothers, New York	540.
Conover Brothers, New York	542.8
Mason & Hamlin, Boston ... A3 Fork,	435.
Mason & Hamlin, Boston	439.
J. H. Foote, New York	449.5
Chas. M. Stieff, Baltimore, Md	456.1
P. Werlin, New Orleans, La	481.8
C. H. W. Ruhe, Pittsburgh, Pa	442 2
Clough & Warren Organ Company, Detroit, Mich	444.9
Hook & Hastings, Boston, Mass	443 8
W. W. Kimball Company, Chicago	451.7
Keller Brothers & Blight, Bridgeport, Conn	458.1
Steinway & Sons, New York	455.9
Krakauer Brothers, New York	454.7
Sohmer & Co., New York	454.5
Hallet, Davis & Co., Boston	453.4
George Jardine & Co., New York	447.5
Kranich & Bach, New York	445.

FIGURE 3.5. Table 1. "Uniform pitch. Piano Manufacturers' Association, New York—Circular No. 10." *The Musical Courier* 23, no. 512 (11 Nov. 1891): 553.

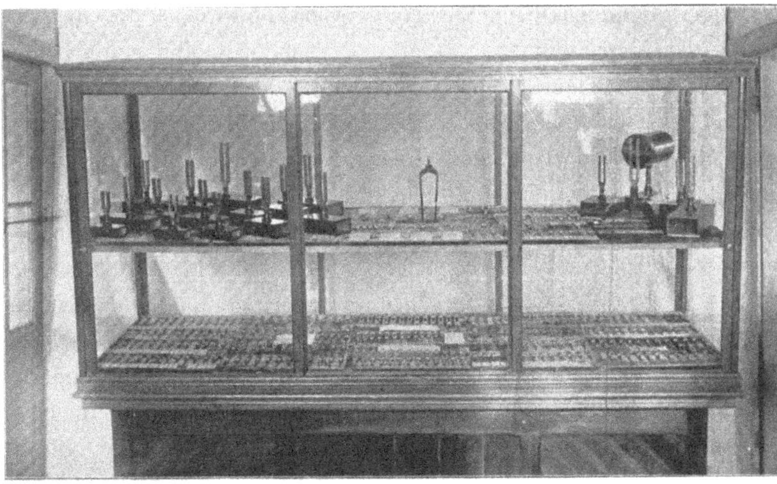

FIGURE 3.6. Frank D. Abbott, *Musical Instruments at the World's Columbian Exposition* (Chicago: Presto Co., 1895), 176.

the dealers in the same [cities] and exhibited them at the N.Y. meeting. It was a perfect explanation and illustration of the dire confusion and want of accuracy of the trade fork question."[81] Featuring tuning forks that produced different sounds, Fuller's display demonstrated audibly the coexistence of different standards on a local scale. In 1893, Fuller took this strategy to a new level, exhibiting his entire general tuning fork collection at the Chicago World's Fair (see fig. 3.6). This display of an overwhelming number of devices materially demonstrated the concept of pitch variation over time and space. Conveying the extent of the problem was a publicity master stroke for the avid pitch reformer. Fuller won an award for his exhibit, with the Fair's judges emphasizing "the historical and unique character of the display" and the "wonderful achievement attained in establishing a universal pitch."[82]

Fuller hoped that a similar strategy of data comparison and demonstration might also persuade the nation's leading piano makers of the utility of adopting the French pitch. Before the piano makers' vote on 6 November 1891, he deployed the same rhetorical technique utilized by Lissajous before the Société d'Encouragement pour l'Industrie Nationale in 1855.[83] Displaying several forks from his collection, both visually and sonically, Fuller delivered a vibrant discourse that merged commercial and artistic arguments in favor of the diapason normal. "In order that we may understand the relation which these various forks have to one another," explained Fuller, "I will sound the standard fork recommended by the committee . . . . The next fork that I will call to your attention is that of Mozart . . . , which I now sound."[84] He then had the piano builders listen to tuning forks associated with other heroic figures from the history of music, namely Handel, Beethoven, and Wagner. "I think you will agree with me when I say that the persons just named are fair representatives of classical and modern composers. Their writings are based in all probability upon the standards here given," asserted Fuller, before asking his audience to consider if there had "been a following of the masters, or has there been a departure . . . ?" The answer to this rhetorical question was clear: the great composers, dead or alive, had worked with lower pitches than the ones currently in use in the United States.[85]

Fuller did not just want to convince American piano makers to adopt the diapason normal: he wanted to help implement it. This required the mass dissemination of accurate tuning devices across the country's factories. Convinced that American standards of sonic precision were not sufficiently developed to guarantee the stability of the standard, Fuller, like Upham before him, took a four-week tour of Britain, Germany, and France. Here, he wished "to pick up what information [he could] on the subject of Pitch, and make certain that the forks are the best possible."[86] During his visit, Fuller chose two manufacturers for America's new standardized forks: Koenig, who made an enormous impression on Fuller when they met in Paris,[87] and Valentine and Carr, a maker from Sheffield that Alexander Ellis and Alfred Hipkins had recommended to Fuller.[88] Koenig agreed to provide American piano factories with "standard forks" of seven inches mounted on resonators, while Valentine and Carr promised to sell forks in much greater quantity, and for a much lower price than Koenig's.[89]

Although essential to the success of Fuller's enterprise, the importation of tuning forks from overseas proved immensely challenging. Time was the manufacturer's main enemy. First, it took months to produce precisely tuned forks. Fuller was well aware of this practical difficulty, having learned during his inspection of Koenig's workshop in Paris of the practices that could provide this quality control. Fuller learned that to ensure the precision of newly produced forks, it was advisable to test the instruments three months after their construction to make sure they had maintained their intended frequency. As Fuller put it, "steel will not settle without it has time [*sic*], that is a law of nature, it is something over and beyond the question of temperature—Molecular change."[90] Confident from the start of his campaign that there was "a general leaning toward Diapason Normal,"[91] Fuller had demanded that Valentine and Carr follow exacting procedures to guarantee the greatest accuracy,[92] and had worked quickly even before the piano manufacturers' decision to establish a supply of European forks—so as to seize the initiative and avoid any delay between the Association's decision and the arrival of the first tuning devices. By December 1891, Fuller had an order for no fewer than 2,500 forks placed with Valentine and Carr, with an initial delivery of 200 of these instruments dispatched to the United States soon after. But there were multiple problems to overcome besides the demand Fuller made for quality control by testing over time—from legal questions of copyright regarding the design of the fork,[93] to custom regulations,[94] to miscommunication with Valentine and Carr.[95] Worse was to come. On receiving the first order of forks on 31 December 1891, Fuller was disturbed to find that the saltwater and moisture of the oceanic voyage had caused the devices to rust and, as a result, lose their blue color.[96] To overcome this difficulty, Valentine and Carr covered the forks of their subsequent deliveries in Vaseline before wrapping them individually in paper.[97]

To avoid such dependency on Europe and resulting challenges arising from transatlantic voyage, Fuller planned to create a robust distribution network by identifying sellers who would stock the new devices,[98] and, more importantly, by arranging for the manufacture of American standard tuning forks on the model of Valentine and Carr's instruments. To this end, Fuller sent copies of the manufacturers' fork to a Boston maker, and instructed him to use a Koenig standard in the construction of these devices.[99] Fuller also conceived of a new method of mass production. Whereas "heretofore fine tuning-forks designated to give absolutely-correct pitch have been invariably made by forging them from a single bar of steel," requiring artisanal skills,[100] Fuller's technique made it possible to assemble several pieces of metal and thus mechanize the production of precision forks.[101] By applying mechanized labor to the production of forks and cultivating an extensive network of distributors, Fuller built a powerful system to support the unification of the nation's pitch.

Fuller successfully convinced the representatives of the music industry to adopt the French standard. But as was so often the case, the piano manufacturers' decision provoked objections from a variety of interested parties. While it helped piano makers, the standard interfered with the activities of other manufacturers of musical technologies. Crucially, it did not suit other instrument players and makers, or smaller businessmen who wanted to operate on a national level and fit American audiences' alleged preference for higher standards. Small piano makers were reluctant to follow suit,[102] as were makers of wind instruments who were in the habit of using British wind bands' "high pitch" of approximately 452 vibrations,[103] and committed to the contemporary perception that a higher sound was a brighter sound. Furthermore, unlike piano manufacture, the market for wood instruments was dominated by European makers. A unifying measure that might stimulate exports was therefore of little interest to American piano builders.

Instrument makers were not the only ones to resist pitch standardization. Individual performers and musical bands also rejected the use of what they saw as a problematic imposition of European authority on their practices. That the question of pitch remained mainly unresolved in the United States in the first years of the twentieth century becomes obvious in the archives of the American Federation of Musicians (AFM). This organization decided in favor of *A* 435 in 1902, yet delegates at each of its subsequent annual national meetings continued to discuss the matter as they struggled to implement the norm at a local level.[104] In 1903, the preparation of St. Louis's monumental World's Fair celebrating the centennial of the 1803 Louisiana Purchase was marked by a revealing controversy when the bureau in charge of the event's musical matters issued a rule imposing the diapason normal. The *Music Trade Review* reported that

> Geo. W. Stewart, manager of the Bureau of Music at the St. Louis World's Fair, announces that the French low pitch has been adopted as the standard pitch for brass bands at the fair. The pitch is a half tone lower than that ordinarily used by the majority of bands. . . . It was selected because it represents modern progress in music, and because American music pitched a half tone higher than European music often injures the work of European singers, whose voices are unaccustomed to high singing.[105]

Several leading bandmasters immediately contested the Bureau's decision. Oscar Milton Wilhite, the leader of the Second Regiment Band of Emporia, in Kansas, "asked Mr. Stewart to name a single band in the West which could compete under this rule. He couldn't name one." Wilhite concluded that the effect of imposing this low tone was to "practically bar amateur bands from the contest," as the cost of replacing their instruments would be prohibitive. He demanded what good contests could do "if they do not encourage the amateur?"[106] While the organizers of the Fair were eager to align with European musical standards, the reaction of the bands revealed the desire of many American artists for a non-European-dominated arts culture.

The hard-won peace that followed the divisive experience of the American Civil War brought with it an unprecedented national expansion of industry, commerce, and the extraction of resources. By the turn of the century, the United States was looking to exert its influence beyond its borders, both diplomatically and through military expeditions. During the 1890s the country developed an imperialist foreign policy in the Caribbean and the Pacific, going to war with Spain, annexing Hawaii, and planning the opening of a canal across Central America.[107] In addition to these political and military programs, the United States started to develop an expansionist cultural policy at the turn of the century. Once the mostly passive recipient of the diapason normal and of European acoustic knowledge and apparatuses, after World War I the United States would impose itself as the dominant voice in pitch debates. A clear indication of this reconfiguration of transatlantic power relations was the adoption in 1917 by the AFM of a pitch standard that was *not* the diapason normal, but rather an *A* of 440 vibrations. This shift from the French norm was barely audible, but it forever changed the course of pitch negotiations.

HELMHOLTZ AT THE CIRCUS: DEAGAN'S CRUSADE FOR *A* 440

The Federation's 1917 move was all the more staggering if one considers the geopolitical context. In 1917, *A* 440 was perceived in the United States as a German standard, yet in April of that year the United States had declared war on Germany. Anti-German sentiment was on the rise, and several periodicals highlighted the unfortunate political connotations of such a pitch selection.[108] In 1919, *The Sun* published an article entitled "Why the German Pitch?" that reminded readers of the 1834 adoption of *A* 440 by the Stuttgart Naturforscher-Versammlung and demanded to know why such a standard had been chosen.[109] A few months earlier, William Braid White had summed up the sense of mystification surrounding the AFM's choice in the journal of the leading union of piano tuners, observing that: "The 440 pitch was originated in the enemy's country, while the international standard is of French origin, and has been officially adopted by piano makers both in this country and in Great Britain. It is, in short, an Allied pitch."[110] Such a comment was rare in *The Music Trade Review*, which typically avoided political subjects.

Both economic and cultural reasons played a role in the AFM's shift from *A* 435 to *A* 440. The United States was the world's largest economy and carried weight within an increasingly global economy. America's music was characterized by the commercial success of blues, ragtime, two-step, foxtrot, and jazz. As popular musical genres began to dominate the musical economy, they acquired unprecedented authority in the pitch negotiations. The gravitational center of pitch conversations shifted from classical, written music, and a valuation of the voice, toward new instrumental genres sustained by the rapid development of the recording and broadcasting industries.

In the world of American music at the turn of the twentieth century, marked by the success of vaudeville and minstrel shows, commercial considerations drove the quest for standardization, but within a completely renewed aesthetic and material context. The industrialization of the musical market and the rise of the entertainment industry not only increased the urgency for pitch standardization, but produced novel representations of sound—such as new definitions of noise.[111]

The principal advocate of the AFM's decision was John C. Deagan, a maker of percussion instruments responsible for the introduction of marimbas to Western musical practice and the development of a new type of xylophone that was to become the standard throughout Europe and the United States.[112] How did this industrialist become involved in the tuning and pitch debates? And why did he choose *A* 440 rather than the French standard favored by piano manufacturers? Answering these questions requires examination of a figure who has remained unknown outside the realms of percussionists and their devotees, despite his significance in the broader history of music.[113] Deagan's activities are exemplary of the centrality of science in instrument makers' careers at the turn of the century, from both a technological and a commercial point of view.[114] Deagan conscripted cutting-edge acoustic knowledge in his building practices and cultivated an image as a scientific instrument maker in order to distinguish himself in the fast-growing market of American percussion making that was sustained largely by the rapid development of the entertainment industry. The history of Deagan's crusade for *A* 440 also shows the interconnection of sound knowledge and the flourishing new businesses of late nineteenth- and early twentieth-century American popular music.

The beginning of the twentieth century was an auspicious moment for percussion instrument makers. They benefited from a new taste for their products, a demand that developed in the worlds of classical and popular music alike, stimulated by the display of non-Western instruments at World's Fairs and the growth of an entertainment industry calling for exotic or simply new sound effects. From 1870, the success of vaudeville offered many opportunities for percussion players and makers. Drums, bells, and other "novelty percussions" became ubiquitous sound accessories in theaters long before they began crowding the shelves of record shops.[115] The importance of percussion in this new American musical soundscape is particularly obvious in the first sound cartoons released by Disney in the late 1920s,[116] whose soundtracks included extensive percussion soli. The relationship between percussion sounds and the entertainment industry would blossom further with the development of television.[117]

The rise of percussion in American musical culture went hand in hand with its manufacturers' appropriation of non-Western instruments, a practice in which Deagan was especially active. Besides introducing marimbas to the United States, Deagan was celebrated as the inventor of the "cathedral chimes"—a set of tubes of different diameters, lengths, and pitches that was used extensively in theaters to imitate church bells, and an instrument that was in fact inspired by the *anklung*, an Indonesian instrument belonging to gamelan ensembles.[118] Deagan was not interested in simply reproducing non-Western artifacts, but in musicalizing, rationalizing, and "civilizing" them. Having commercialized bamboo chimes in 1897, he developed a metal version of the instrument, which he claimed enhanced its technical quality. The trade brochures issued by Deagan Inc. promoted the firm's leading products: bells, xylophones, chimes, and marimbas that were "improved" versions of either popular, old, or "savage" objects. Innovative alterations were portrayed as the result of taking a scientific approach to percussion making within a factory-laboratory under the authority of a skilled builder-acoustician, namely Deagan himself.[119]

These efforts to rationalize the allegedly uncivilized, unscientific, and unmusical native instruments were manifest in the increased regulation and refinement of the primary materials used in their production. "The manufacture of Deagan xylophones commences with the selection of the wood in the tropical forests. Only young trees of straight and even grain are chosen. After the bars have been cut to size in the Deagan factory the stock is thoroughly seasoned for a number of years,"[120] claimed a 1920 trade catalog, which then went on to detail how this attempt to "improve" percussion instruments involved the transforming of extracted materials into instruments of extremely precise pitch. Listening to one of the early Disney sound cartoons gives a good sense of the achievements percussion makers made in the first decades of the twentieth century. *The Haunted House* (1929), especially, displays a series of virtuosic scales performed on perfectly tuned xylophones, demonstrating the total integration of these instruments into an increasingly uniform Western musical soundscape.

Deagan's efforts to control the tuning of percussion instruments drew him into debates over pitch standardization. Although he was eager to portray his contribution to these conversations as an effort to rationalize musical practices according to mathematical principles, Deagan's motives were also commercial. The tuning of percussion instruments cannot be adjusted, yet Deagan's products were intended to travel all over the United States and beyond. Thus, his interest in pitch standardization was not merely philosophical, but every bit as material as that of America's piano manufacturers. The question, then, is why did he not simply continue the efforts of piano builders in promoting the international standard adopted in Vienna, given his ambitions to conquer European markets? In the absence of sufficient archival evidence, this is hard to answer conclusively, but it was almost certainly the result of industrial concerns, linked to the development of musical entertainment in America, combined with the very different aesthetic and cultural assumptions that characterized nineteenth-century German and British acoustics.

In 1916, Deagan patented the "Dea-gan-ometer," an invention he claimed would "solve . . . the mystery of pitch." As Deagan explained,

> One of the greatest problems confronting conscientious orchestra conductors and musicians today and one causing them no inconsiderable amount of worry and anxiety is the old question of pitch and tuning . . . . The various nations have not been able to agree on the question of a universal pitch . . . . Recognizing this unfortunate condition we have produced a new standard tone measure, the Dea-gan-ometer, the function of which is to provide an easily understood, easily accessible and irrefutable PITCH STANDARD for the musician to work from, thus eliminating all uncertainty and guesswork regarding pitch or tuning. When once used, the Dea-gan-ometer will be found as indispensable to the musician as the thermometer is to the physician, the compass to the mariner or the rule or scale is to the artisan or merchant.[121]

Consisting of four metal bars tuned to pitches separated by only one vibration, the instrument clearly took its inspiration from Scheibler's early nineteenth-century tonometer and subsequent descendants.[122] It also recalled Rudolph Koenig's clock fork, combining this apparatus with a pendulum attached to the frame and providing the user with an easily observable indication of seconds to help with the counting of beats.[123] The Dea-gan-ometer finally drew directly on Deagan's 1890s experimentation with the tuning of chimes—it was made of a set of metal tubes resembling these earlier instruments (see fig. 3.7).[124]

During 1916–1917, Deagan published several booklets on the topic of tun-

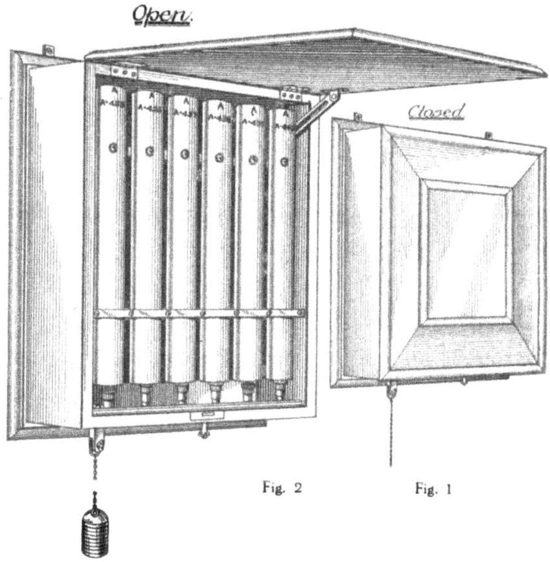

Fig. 2 Fig. 1

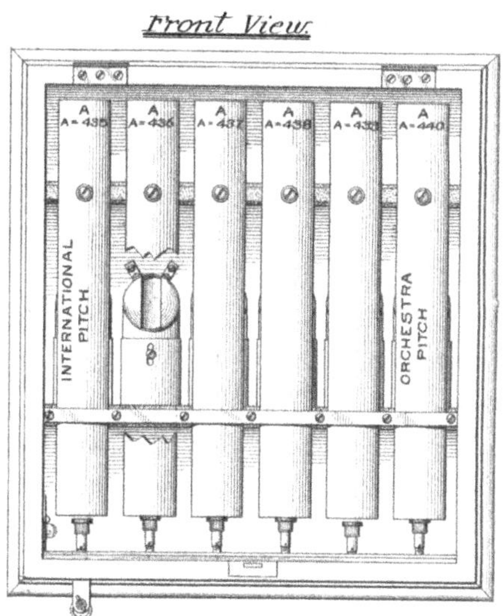

FIG. 3

Fig. 3—Shows the six bars of the Dea-gan-ometer, with the number of vibrations stamped on each—the Master Tuner A=435, International Pitch, is guaranteed absolutely correct. All the other bars are also correct, as any musician can demonstrate by the pen-dulum and the beats. Bar A=436 is cut away to show the resonator in the back of each bar.

FIGURE 3.7. Deagan, *The Dea-Gan-Ometer*, figures 1, 2, and 3 (4; 6).

ing, which he sent to the Library of Congress, where most remain today.[125] These documents demonstrate Deagan's mastery of acoustical matters and include an especially imposing table of the chromatic scale's frequencies for *A* 440 in equal temperament.[126] For someone so eager to embrace Helmholtz's legacy, the celebration of equal temperament seems initially dissonant.[127] Deagan, however, invoked the authority of Johann Sebastian Bach to justify this measure. He also claimed to be the first to finally put into practice the celebrated composer's theory of sound. Of course, equal temperament was in fact an economically attractive option, allowing builders to mass-produce the components of their products. Nevertheless, Deagan's choice of *A* 440 actually did resonate with Helmholtz's writings. The German physicist never referred to the French standard of *A* 435, although its adoption was being discussed in several parts of Germany at the time he was writing *Die Lehre*. Helmholtz, however, always used 440, the Scheibler pitch, as a point of reference. These two different norms were drawing on different intellectual traditions: their amalgamation epitomized Deagan's project of applying Helmholtz's acoustics to the business of musical entertainment.[128]

More obvious an influence on Deagan than Helmholtz, however, were Ellis's writings. Indeed, Deagan explained that he used the ratio 1.0594631 to calculate the various pitches of the tempered scale, a number that Ellis provided in the "Appendix" to his translation of *Tonempfindungen*.[129] Furthermore, Deagan's *R Catalog* includes an "Outline History of Musical Pitch During the Past 200 Years" that is almost exactly identical to the one Ellis gave in his 1880 paper on the history of musical pitch, the only difference being Deagan's inclusion of the AFM's 1917 resolution on pitch (see fig. 3.8). In Deagan's outline, Beethoven is also presented as the father of *A* 440.

Deagan borrowed from Ellis without acknowledgment, but did not follow the British acoustician's recommendation of the French pitch. His justification for choosing 440 actually invoked the argument of another figure prominent within nineteenth-century acoustics: the piano builder A. J. Hipkins, who had been Ellis's collaborator. As shown in the previous chapter, in 1896, Hipkins had suggested raising the pitch from 435 to 440 in order to account for the usual temperature of concert halls in England and, in the process, raised a whole new range of problems within the field of pitch negotiations. "In changing the frequency of International Pitch vibrations from A=435 to A=440 the American Federation of Musicians has not really changed the pitch at all," Deagan wrote in an article defending the decision of the Federation within the context of debates over America's new standard that were becoming increasingly awkward, in terms of both cost and political impli-

OUTLINE HISTORY OF MUSICAL PITCH DURING THE PAST 200 YEARS

Semi-tones and tenths of Semi-tones	Pitch Vibrations	Where and by Whom Used-	Name of Pitch
S.	A		
0.0	370	Lowest, zero-point.	Church Pitch lowest
0.2	374	Hospice Comtesse, 1700.	
0.3	377	Schlick, low, 1511; Bedos, 1766.	
1.0	392	Euler's Clavichord, 1739.	Church Pitch low
1.1	395	R. Smith, 1759; Roman pitch pipes, 1720.	
1.2	396	De Caus, 1615; Versailles Chapelle, 1789	
1.4	403	Mersenne Spinet, 1648.	Chamber Pitch low
1.6	407	Sauveur, 1713.	
1.7	408	Mattheson, Hamburg, 1762.	
1.7	409	Pascal Taskin, court tuner, 1783.	
2.0	415	Dresden chained fork, 1722.	European Mean Pitch for two centuries
2.2	420	Freiberg, 1714; Seville, 1785.	
2.3	422	Mozart, 1780. J. S. Bach, 1750.	
2.3	423	Handel, 1751.	
2.4	424	Praetorius's suitable pitch, 1619; original Philharmonic, 1813.	
2.5	428	R. Harris, 1696; Opera Comique, 1823.	
2.7	433	Sir George Smart's fork, 1820-26.	Compromise Pitch
2.8	435	FRENCH DIAPASON NORMAL, 1859	
3.0	439	NEW ENGLISH PHILHARMONIC, 1896.	Modern Orchestral Pitch, and Ancient Medium Church Pitch
3.0	440	BEETHOVEN, 1816-27.	
3.0	440	SCHEIBLER'S Stuttgart Standard, 1834.	
3.0	440	OFFICIAL PITCH, Am. Federation of Musicians, 1917.	
3.1	442	Bernhardt Schmidt, low, 1690.	
3.2	445	Madrid, 1858; San Carlo, Naples, 1857.	
3.2	446	Broadwood's Medium, 1849; French Opera, 1856; Griesbach's A, 1860.	
3.4	449	-C 534; Griesbach's C 528; 1860.	
3.5	451	Lille Opera, 1848; British and Belgian Army, 1879.	
3.5	453	Mean Philharmonic, 1846-54.	
3.5	454	English high, 1848.	
3.6	455	Highest Philharmonic, 1874; Broadwood, Erard, and (English) Steinway, 1879.	
3.6	456	Vienna, high, 1859.	
3.7	457	(American) Steinway, 1879.	
3.8	458	Great Franciscan Organ. Vienna, 1640.	Church Pitch high
3.9	460.9	American high, (Vaudeville) 1880.	
4.0	466		
4.3	474	Tomkins, 1668; B. Schmidt, high, 1683.	
4.5	481	St. Catherine's, Hamburg, 1543.	
4.8	489	St. James's, Hamburg, 1688.	
5.0	494	St. James's, Hamburg, 1879.	Church Pitch highest
5.1	496	Rendsburg, 1668.	
5.3	504	Schlick, high, 1511; Mersenne, ton de chapelle, 1636.	
5.4	506	Halberstadt, Cathedral, 1361.	
6.0	523	Cassel, Germany, 15th Century.	
7.0	454	Mersenne, ton de chambre, 1636.	Chamber Pitch highest
7.3	563		
7.4	567	Praetorius, North German, very old.	

J · C · DEAGAN · INC · CHICAGO · U · S · A

FIGURE 3.8. J. C. Deagan Inc., *Catalog* R [1920]. Courtesy James A. Strain and the Percussive Arts Society.

cations. In the same article, he explained that at the normal temperature of theaters and other music venues, most brass instruments' pitches tuned to 435 at 15°C would reach 440 vibrations.[130] Deagan almost certainly knew of Hipkins's article, given that he claimed to have researched the question extensively, not by measuring old instruments, but by reading the latest literature on the subject.

His temperature argument was in no way original, but Deagan's justification for *A* 440 still introduced a completely new perspective to the debates. Rather than mentioning classical musical practices, as previous negotiators had invariably done, the builder turned to the world of popular music to vindicate his selection. Deagan made this point explicit through the testimony of a clarinet player, as well as by citing the tuning practices of popular bands. Mobilizing the experiences of musicians, he described how

> As it is the clarinet (or the oboe where used) to which the rest of the instruments are usually tuned, the pitch of the clarinet (or the oboe) is generally the pitch of the entire band or orchestra.... Piano tuners... who have tuned pianos to 435 for orchestra use have almost invariably been assailed with the cry from clarinet players that the pitch was too low.[131]

Note here Deagan's decision to include references to classical orchestras only in parenthesis, reflecting his emphasis on popular music. Deagan's *A* 440 was very much fashioned as a modern standard for a modern age.

The change in standards in 1917 marked a crucial point in the history of pitch negotiations. By championing *A* 440, Deagan was refusing to adopt the current European norm and imposing his own cultural values and economic interests on the international stage. As well as contributing to pitch debates through publications, Deagan promoted the standard with a number of national organizations. When the AFM met for their annual meeting in 1918, the president of the Chicago section "saw to it that each of the hundreds of delegates carried home..., with the compliments of the Chicago Federation of Musicians, a Deagan 'A=440 Tuning Fork' and a Deagan A=440 Studio Wall Chart."[132] A few months later, Deagan sent a "Master Tuner"[133] to the American Bureau of Standards to help implement the nation's new measure. He also urged the Bureau to "take up and settle the question of a uniform pitch to be used by musical instrument manufacturers and musicians of the entire world—a world in which I believe this instrument [i.e., the 'Master Tuner'] will be found to be highly useful."[134] The head of the Bureau thanked him and explained that they would take time to think about the issue of pitch once the war was over.

Deagan's intervention in American debates over pitch unification prompted the reopening of negotiations between representatives of the music industry as a whole. Although at odds with the interests of piano and organ builders, Deagan's reform would ultimately triumph on a national stage, attesting to the new cultural authority of increasingly diverse instrument makers and the country's prosperous popular music industries. It also showed a rising awareness of the new economic weight of the American domestic music market and an assertion of confidence from the United States' musical networks.

"THE UNITED MUSIC INDUSTRIES"

The controversy surrounding Deagan's pitch went well beyond nationalist and diplomatic concern over German influence. Piano and organ builders who had adopted *A* 435 as a standard were especially eager to contest the validity of his reform. The relative consensus reached in 1891 after Fuller's campaign had been immensely difficult to attain, so it was only natural that many feared Deagan's move would jeopardize all hope of the United States ever reaching a lasting agreement on sonic matters. Far from creating chaos, however, Deagan's intervention prompted negotiations that eventually resulted in an unprecedented international concord.

In 1925, the American Music Industries Chamber of Commerce appointed a Technical Committee on Standard Pitch. The Chamber was in fact an amalgamation of several associations, including the Piano Manufacturers' Association,[135] one of the few national musical associations in the United States at the time, along with the American Guild of Organists, the National Association of Organists, the Music Teachers' National Association, and the Music Supervisors' National Conference.[136] The president of the National Association of Piano Tuners, Charles Deutschmann, chaired the pitch committee, which included piano magnates Henry Ziegler and John Anderson, respectively directors of Steinway & Sons and Chickering & Sons. The nation's piano-making interests were also represented by William Braid White, as well as by the leaders of the firms Lyon & Healy and C. Bruno & Sons, which built diverse instruments. Mathias Peter Møller, a successful Danish immigrant settled in Maryland, and Ernest Skinner, from Boston, represented America's organ builders. The head of the flourishing firm Conn spoke on behalf of wind instrument makers, while the director of the Chicago firm Gulbransen offered the insights of a player piano manufacturer. This powerful network of industrialists, all successful owners of the United States' premier music instrument businesses, mobilized a great variety of technological expertise.

The Technical Committee's commitment to employing the very latest scientific knowledge of sound was reflected in the appointment of Dayton Clarence Miller, a professor of applied science at Case Western University. An authority on sonic matters, Miller had promoted the implementation of the international pitch since the 1910s, having checked standard forks for both the Bureau of Standards and the most prominent piano builders of the country.[137] Throughout the duration of the committee's investigation, Miller offered "the use of his laboratory and equipment at Case," an invaluable gesture according to *The Music Trade*, which observed that "this laboratory is one of the best equipped in the country for the study of sound and tone production."[138]

Invested with the authority to represent the American music industry as a whole, the Music Industries' Technical Committee was charged with examining Deagan's German pitch. Chairman Deutschmann claimed that the growth of amateur musical culture in the United States made "establishing a standard pitch . . . [a] vital necessity."[139] Eager to avoid the divisions that had plagued Deagan's recent reform, the inquiry distributed a questionnaire to all members of the National Piano Technicians Association that asked about both the pitch they favored and how the piano factories operated. In gathering this information, the committee was attempting to gain a sense of what sort of discipline could be undertaken to ensure the stability of the standard within the factories. The questionnaire took respondents through the stages of tuning a piano, starting with "chipping" (a sort of initial rough tuning of the piano strings), continuing on to the main "tuning," and ending with "the later tunings." After technicians had completed these forms, factory superintendents specified the ideal number of times these actions should be repeated, and how long the waiting periods between tunings should be.[140]

According to White, who kept his readers informed of the committee's deliberations through his journal, *The Trade Review*, the results of this survey were unequivocal. Drawing upon the data gathered through the questionnaires, White explained to his readers that "approximately 98 per cent of the manufacturers of fixed pitch musical instruments have found themselves compelled during the last five years to adopt a pitch equivalent to 440 double vibrations per second, for the A above middle C."[141] In response to this finding, White was forced to admit that although he had initially been skeptical about Deagan's reform, his position partly resulted from a misconception. The committee's national survey left little doubt that *A* 440 was now the only practical measure that could bring order to America's musical instrument industry. Conceding that Deagan's unit was indeed authoritative, White wrote:

> The committee recommends that the 440 A be adopted as the standard pitch used by all manufacturers of musical instruments under the jurisdiction of the M.I.C.C. It recommends that the new pitch be called the American Standard Musical Pitch, adopted by the Music Industries Chamber of Commerce 1925.[142]

White applauded what he saw as a remarkable sign of solidarity among American manufacturers, who were finally seeing "that their common interests are far more important and numerous than their common jealousies and enmities."[143] Summing up this unprecedented consensus, he declared this to be the age of the "United Music Industries."[144] Of course, this was a somewhat rose-tinted evaluation. A consensus may have been reached across businesses of similar sizes that pursued shared commercial objectives. But for the country's smaller manufacturers, especially the piano and organ builders who had implemented the previous standard of *A* 435, the committee's adjudication presented a stark choice between adopting the new pitch of *A* 440 or facing exclusion from what was an increasingly integrated market.

As satisfying a consensus as this survey was, it was only a first step on the road toward acoustic uniformity. Once again, the main concern was implementation. As European countries had done with Koenig and the Viennese Bureau, the committee tasked Miller with regulating America's standards from within his laboratory at Case Western. In this way, the university joined the growing network of centers for controlling musical pitch, from the Paris Conservatory and University La Sapienza in Rome, to the University of Vienna and the Physikalisch-Technische Reichsanstalt in Berlin.[145] Miller was an astronomer who led experiments on ether drift. He was also a controversial opponent of Einstein's theory of relativity. He brought to the study of sonic phenomena the same extreme accuracy that was so central to astronomy. In 1908, Miller contributed to a long tradition of sound visualization by inventing the phonodeik, an instrument that allowed for the photographic record of the shape of complex sound waves. Miller used this to popularize acoustics, giving spectacular lectures throughout the United States.[146] His experiments with speech were particularly appealing; and in 1917, newspapers reported that he mesmerized audiences by providing them with a graphic representation of a kiss.[147] In addition, Miller was an accomplished flutist who owned a collection of more than 1,650 flutes. He also composed several musical works in various genres.[148]

To calculate the frequency of tuning forks, Miller used a device of extreme acoustic precision: Koenig's clock fork. In using this apparatus, Miller followed Koenig's recommendations by returning to those who submitted their fork for examination a "certificate of exactitude" on which he detailed how "the exact temperature of the fork . . . observed at each measurement, and the observations . . . reduced to the standard temperature coefficient."[149] Further, Miller combined the clock fork with his own phonodeik to visualize the vibrations produced by the forks he measured. White boasted that

> When these forks are completed the music industries of the United States will be in possession of the most accurate set of standards for pitch which has ever been produced. It will be possible to attain certified duplicates of these forks, it is announced by the committee, for all purposes of musical instrument manufacturing without the slightest difficulty and at very small expense.[150]

In other words, the United States, once the recipient of European standards of acoustic precision, was taking an authoritative lead in the verification of tuning forks and standardization of pitch.

The committee was eager to address broader problems of implementation beyond the accuracy of tuning forks. Most importantly, they considered that the success of their reform would depend on the degree to which factory workers, from superintendents to employees, could be appropriately trained. From the evidence they gathered through the questionnaire, they concluded that the reason past efforts to establish an international pitch had failed was due to the insufficient regulation of factories. As White explained:

> In many shops the caprice of the head tuner and the exactness of his own personal forks have been allowed to rule so that there has been confusion worse confounded in this very important matter, and gradually the superintendents and the executive heads of many manufacturing houses have come to neglect the whole matter.[151]

To solve this problem, White thought it necessary to establish a form of legal contract between manufacturers. He suggested the circulation of pledge-forms that would bind the makers morally, if not legally.

CONCLUSION

Standards play an important role in regulating markets.[152] They also have a particularly privileged relationship with organizations: while standardization can help regulate the practices of an organization, organizations are in turn efficient vehicles for the implementation of such ordering measures.[153] The history of musical standardization in the United States illustrates this point. In the final decades of the century, concert pitch became a crucial tool for integrating the American musical market, connecting the business of instrument making with musical practice, from pianos and organs to wind instruments, percussion, and player pianos.

It is not clear whether in its efforts to implement the French diapason normal and, subsequently, the new pitch *A* 440, the United States was or has been more successful than the Europeans. Evidence of local resistance from small-scale artisan manufacturers and amateur bands suggests that beyond the nation's most eminent factories and music venues, a great variety of tuning practices coexisted before the outbreak of World War II. As in Europe, American pitch reforms confronted aesthetic resistance fueled by an appreciation of high, allegedly "bright" sounds. Further research remains to be done into how much musicians appropriated the standard for different musical genres, from classical and Romantic symphonies to folk music, from opera to foxtrot, ragtime, jazz, and pop music.

One way or another, the consensus of instrument makers around *A* 440 facilitated its introduction on a national and, eventually, international level. More than overseas, however, economics rather than aesthetics was the force shaping pitch negotiations. The considerable involvement of actors from the American music industry, beginning in the 1880s, enabled a type of conversation quite new in the history of pitch negotiations, one between interlocutors who, though initially opposed on a number of concerns, shared common values. Importantly, American instrument makers were aware as never before of the global implications of securing a standard for commerce. All in all, the transatlantic history of musical standardization shows how industrial capitalism reshaped the sounds of Western music cultures, all the while demonstrating the power of culture and aesthetics to fine-tune political economies—a situation nowhere more audible than in the United States' shift from *A* 435 to *A* 440 as a musical standard.

The efforts of the United States in the late nineteenth century to develop its own influence within classical music had by the late 1930s established an international standard that would be followed by the rest of the world. The question of what a "distinctly American sound" could be,[154] a question prominent within the country's musical circles at the end of the nineteenth century, had found in the science of acoustics and the workings of cultural economics a far more definitive answer than those suggested by the existing musicological scholarship that focuses on composers, genres, and repertoires.

The adoption of *A* 440 not only had implications on a national level. In 1935, the American Bureau of Standards started broadcasting a signal at a frequency of 440 hertz. This prompted the reopening of pitch negotiations in Europe. Besides investing the American standard with an official character, the broadcast of the *A* 440 signal revealed the potential of new sound media to unify soundscapes on a global scale. If radio was a crucial tool for standardizing pitch, then it is equally apparent that broadcasters considered pitch standardization a key condition for the creation of a unified radiophonic space. Having examined the laborious efforts to regulate America's industrial factories during the early twentieth century, I now turn to the subsequent work of broadcasters to discipline the way music was performed and listened to around the world.

CHAPTER FOUR

"Pitch in Our Time"

International Concord and the Engineering of an Interwar Standard

> Sometimes the technician has more of an international spirit than the state representative.
>
> ROBERT DE TRAZ

The year 1939 is not usually thought of as a time of international concord.[1] Yet this is when Western countries finally agreed upon a common standard pitch. How can one explain this surprising turn of events? And what did broadcasting and electroacoustics have to do with it? In answering these questions, this chapter highlights the unsuspected role of radio as an actor of interwar diplomacy while showing how concert pitch became a harbinger of peace in the troubled years leading up to the Second World War.

On 5 October 1937, the foreign director of the BBC forwarded to the head of the broadcaster's Overseas and Engineering Information Department, L. W. Hayes, a letter from Gustav Schwaiger, chief engineer at the Austrian Radio Company and a member of his country's commission for the verification of tuning forks. Following up on the recommendation of the Comité International Acoustique (CIA), a body created only a few months earlier in order to enhance international cooperation in the field of electroacoustics, Schwaiger explained that the Austrian commission was collecting data on tuning practices from across Europe and intended to initiate a new international conference on musical pitch. Schwaiger wanted to know if there were institutions in the United Kingdom that had any degree of authority over this question and with whom he could get in contact. Along with the Austrian engineer's letter, the BBC director of foreign affairs included a memo for Hayes, wondering if he could "look at this letter in the first place as it comes from Dr. Schwaiger and mentions the Comité International Acoustique, which is usually your affair? No doubt you will wish to get in touch with Music Department. Is this a question which can be left to the Comité International Acoustique without the intervention of the artistic side?"[2] Expressing surprise, Hayes responded that he did not "know quite why this came up in the CIA but it is obviously a question needing International action. As I see it, it is a question which has no engineering implications. On the other hand, there are obviously artistic ones."[3]

Some eighteen months later, on 11–12 May 1939, a temporary resolution to the international regulation of pitch was reached in London when, amid rising international tensions, delegates from Britain, France, Italy, Germany, and Holland agreed on the following three points: "That the international standard of concert pitch be based on a frequency of 440 cycles per second for the note A"; "That this value be maintained within the closest limits possible by soloists, orchestras, choirs, etc. throughout all musical performances, and also in recorded music"; and "That with a view to reducing the necessary tolerances to acceptable values, a set of technical recommendations be adopted, preferably on the basis of international co-operation."[4] The speed with which pitch standardization was ratified at this moment—on the eve of World War II—underlines the spectacular empowerment of representatives of the world of electroacoustics in pitch negotiations. At this conference most states were represented by their standards agencies, and their decision was the result of two years of negotiations initiated by the CIA.[5]

The Comité had been founded during the first international conference on acoustics held jointly in Paris during the 1937 World's Fair, from 30 June until 3 July, by the International Standards Association (ISA, forerunner of the International Organization for Standardization, or ISO) and the International Electrotechnical Commission.[6] The participants in the 1937 conference in Paris identified several priorities for international cooperation that reveal the entanglement of sonic knowledge with industrial developments.[7] These included: the creation of an international acoustics vocabulary; the definition of units and scales required for noise measurement; the discussion of appropriate procedures for determining sound absorption coefficients, reverberation time, and sound insulation; the development of medical acoustics; and the standardization of musical pitch.[8]

A STANDARD TO BUILD PEACE

On 1 July 1937, a subcommittee within ISA's acoustics committee (which would operate as ISA Subcommittee 43/3[B]) "unanimously agreed that it was necessary to unify, in the various countries, the number of vibrations of the standard 'pitch.'"[9] Following this resolution, the committee decided that delegates would "send to the German Secretariat within one month the proposals of their National Committees on this subject." In a circular, Martin Grützmacher (1901–1994), the German secretary of the subcommittee, asked countries to submit not only a proposition of a standard but also "regulations regarding the type of control of the concert pitch."[10] Furthermore, the committee was considering whether "to hold a meeting if possible within six months . . . in Vienna." With the international scene worsening, however, this meeting was both delayed and displaced, and finally took place in London in May 1939.

Beyond the connection between musical standardization on the one hand, and interwar industrial and scientific electroacoustic developments on the other, Hayes's 1937 memo reveals how, by the late 1930s, the institutional context of pitch negotiations had changed radically from the early twentieth century. Once the prerogative of states, scientific societies, and musical associations, on the eve of World War II, the establishment of a unified pitch was primarily the product of concerted efforts from sound engineers and broadcasting companies.

Following in Hayes's footsteps, this chapter asks why pitch standardization, if it was an issue with "no engineering" implications, ended up on the program of electro-acousticians and broadcasters.[11] At first sight, this appears to be the result of technical necessity. Yet to reduce the question of international cooperation over pitch to such a technologically deterministic interpretation ignores the complex political, cultural, and ideological concerns at stake during the late 1930s. The introduction of sound engineers and broadcasters into these pitch debates was not only a matter of resolving the practical challenges that the expansion of broadcasting entailed, but represented a combination of technical and ideological considerations.

The involvement of sound engineers was the product of interwar conceptions of music as a harbinger for peace. After the horrors of the First World War, there was a shared belief in Europe that music could help contribute to international harmony. The activities of the International Musical Society, created just before World War I, as well as those of the League of Nations, which developed multiple musical projects, including one dedicated to the unification of musical pitch, all attest to this phenomenon.[12] While the phonograph opened global horizons to musical markets, radio was a powerful instrument in creating international communities of listeners.[13] These new sonic media, with their ability to transcend national borders, particularly intrigued those actors who aimed to foster sociopolitical and cultural integration through music.

In the context of escalating international tensions between 1937 and 1939, the work of broadcasters took on even greater political significance. The political implications of standardizing pitch amid the rapid escalation of international tensions were explicit in the materials prepared for Sir Cecil Graves, head of the BBC, just a few days before he gave the keynote address at the London conference in 1939. These came from the director of the British Standards Institution, Charles Le Maistre, who pulled no punches in his emphasis of pitch's place within the deteriorating international scene:

> In times of political unrest the stabilising effect of such technical conferences, assisting as they do in smoothing out difficulties in the industrial world, are very much to the good. The British Standards Institution, which is a chartered institution with Government, is doing a great deal to bring the spirit of co-operation and mutual concession into industry through its many standardising committees.[14]

The activities of the International Broadcasting Union, or IBU, one of the two international organizations that took part in the 1939 conference in London, exemplify the involvement of broadcasters. The IBU was established in 1925 to organize radio transnationally and to address the challenges raised by the fast development of broadcasting in Europe.[15] It consisted of two main organs: a technical committee handling practical issues regarding cross-border infrastructures, electrical interference, and radio wave propagation; and a program department "aimed to facilitate and develop the use of radio to encourage rapprochement between peoples."[16] Through this dual agenda, the IBU hoped to foster "the idea of a Europe of peacefully co-existing nation-states with differing traditions, political systems, and economic structures."[17]

Standardization was part of broadcasters' efforts to secure peace through the cultivation of music. In December 1937, following ISA Subcommittee 43/3[B]'s initiative to launch an international consultation on pitch standardization, the IBU created a "study group" dedicated to the problem. Raymond Braillard, the French head of the IBU's technical committee, justified this initiative by invoking the "grave inconveniences resulting from [pitch variations] as much for the exchanges of orchestras as for international relays and for all the techniques of recorded music."[18] The IBU saw the cultivation of high-quality live and recorded classical music as a way to promote reconciliation. Throughout the 1930s, the IBU developed several programs to encourage the expansion of orchestral music as a vector of sociopolitical and cultural integration. Considered a universal language, music appeared to be the best medium to communicate transnational cooperation.

The project raised several questions. First of all, what was high-quality music? And should it be live or recorded, with or without words, presented in short or long performances? Technological and legal issues included the development of infrastructures suitable for transmitting the sounds of orchestral music, as well as emerging questions of copyright regarding the transnational circulation of recorded material. The broadcasting of "good" music was a highly political project, seen unanimously as a way of building a pacific supranational community.[19]

Throughout the interwar period, the specter of World War I infused technical and cultural initiatives with the motive of preventing a return to global conflict. Thus pitch, in the "age of electro-acoustics,"[20] remained a diplomatic object, inscribed with romanticized notions of music as a universal language. The months of discussions over pitch reform were marked by growing alarm over the increasingly expansive politics of Nazi Germany. ISA Subcommittee 43/3^B's decision to hold a meeting in London, rather than Vienna or Berlin as initially planned, was a concession to delegates who were reluctant to endorse the antisemitic policies and military expansion of the Third Reich. Despite political tensions, sound engineers at the 1939 conference in London sustained a level of consensus unprecedented in the history of pitch standardization. The transnational networks of electro-acousticians offered efficient and resilient channels of negotiation that enabled countries from across Europe and North America to build a consensus over what had been a long-debated question.[21] At the very moment when the world's powers were sliding toward the catastrophe of world war, these same nations agreed to an unparalleled system of musical integration.[22]

For the sound engineers at broadcasting stations, the standardization of pitch was important in three ways. The first was not new: like previous actors involved in pitch diplomacy before them, interwar broadcasters were eager to secure uniformity to make it easier for musicians to travel and perform at different locations, a sine qua non for the integration of European musical programs.[23] Second, classical music was not easy to transmit through radio waves. In comparison with speech, for instance, orchestral music's frequency spectrum was huge, yet broadcasters aspired to high standards in the quality of its transmissions. The IBU's decision to use cables rather than wireless for the exchange of orchestral music helped remedy these tensions, as did the selection of specific repertoires. The relatively limited frequency spectrum that characterized "Baroque" compositions, for example, made this genre especially suitable for high-quality broadcasting. Concomitantly, it was easier to install a chamber music ensemble in a studio than a monumental philharmonic orchestra.[24] Third, pitch inconsistency aggravated the technical challenge of broadcasting classical music, as the disparity of reference pitches across space indeed increased the spectrum of frequencies transmitted by radio and thus further complicated the work of broadcasters.

In addition to these practical issues, the lack of uniformity in tuning practices was seen as an obstacle from an aesthetic point of view. And aesthetics was key to the political project of broadcasters. As Henri Bonnet and Franz Wilhelm Beidler-Wagner, the two influential experts of the League of Nations who shaped interwar ideas on musical broadcasts, argued, the exchange of classical music programs would help imbue listeners with a sense of community.[25] But while radio could, for the first time, give shape to the concept of music as universal language, broadcasters warned that the inconsistency of reference pitches undermined this concept: pitch variations indeed emphasized national and local disparities. The more homogeneous the musical canon, the more obvious the discrepancies between two performances of the same work became. Building an agreement about pitch was thus crucial to broadcasters' project to construct a consensus among European nations.

As head of the BBC, Sir Cecil Graves emphasized this point in his opening speech to the 1939 London conference, on 11 May. Drawing on well-established rhetoric, he first outlined the advantages of standardization for instrument makers, as well as for singers and instrumentalists, whose voices or strings, he argued, were strained by high pitches.[26] However, Graves then proceeded to go beyond this customary list of beneficiaries of pitch standardization by stressing that audiences too would appreciate such musical uniformity. As Graves put it: "as far as listeners are concerned, whether they listen in the concert hall or by broadcasting, no real comparison on the score of brilliancy of two different performances of the same symphony by different conductors is possible unless the two performances have been to the same pitch."[27]

Graves's observation assigned an important and specific function to the listener within the process of broadcasting. The aesthetic ideals that had governed efforts to unify pitch on a local scale since the beginning of the nineteenth century persisted. For interwar broadcasters, just as for members of France's 1824 attempt to regulate pitch at the Paris Opera, pitch was not only a matter of intonation, but a constitutive part of tone color and orchestras' aesthetics. It was a defining element of timbre. Standardizers considered pitch to be a key factor for the creation of a homogeneous musical experience within a given territory.

Whether nineteenth-century concertgoers could actually distinguish between the various tones that Parisian orchestras employed remains open to historical debate, but new radio technology in the 1930s appeared to allow audiences to hear back-to-back performances from all over Europe, making the experience of discerning pitch discrepancies, and thus differences in timbre, a realistic proposition. For radio engineers, pitch uniformity was a precondition for the constitution of a truly international musical space not only from a technical, but also from an aesthetic point of view. In the words of acoustician and member of the British delegation at the 1939 conference Llewelyn S. Lloyd, "as soon as broadcasting entered the field of discussion, a new background was automatically provided. It became impossible to cling to any purely insular views, for broadcasting afforded a ready means of comparing the pitches used in practice in different countries."[28]

The definition of a normal pitch was therefore no longer purely a technical issue, but part of a broader cultural agenda in which technicians worked toward increased international integration. They felt their work was both noble and socially important. If they could solve the seemingly intractable problem of sound variations, sound engineers would literally contribute to cultivating peace through the cultural integration of Europe, and therefore to a pacific future. Graves's address illustrated the politics at stake. He concluded his speech by asserting that "it is no small achievement for the representatives of many nations to come together and discuss a subject that is very much bound up with the *emotional factor*. General agreement on principles and practices might well be a promising portent, in these days of emotional stress."[29] In this way, Graves explicitly designated the international context as justification for broadcasters' project to tune the world. To fully recover the meaning of these words, remember that by May 1939, the prospect of a new global conflict was becoming more certain day by day. Following Hitler's annexing of Austria and occupation of Czechoslovakia, Poland was preparing to face a German invasion; a month earlier, Mussolini had begun to proceed with his project to conquer the Mediterranean with the invasion of Albania;

and on 11 May, the day on which Graves delivered his address, the Japanese and the Soviets went to war over the Khalkhin Gol, the border between Manchuria and Mongolia.

A NETWORK OF SOUND ENGINEERS

Beyond the mix of ideological and technical concerns of engineers motivating the acoustical cooperation of the late 1930s lay a third factor: the proliferation of international networks of engineers, itself the result of competing nationalist agendas. Existing historical studies have considered the institutionalization of acoustics in the frame of specific national contexts. This process, however, was part of transnational conversations about pitch. By the end of the 1930s, exchanges within the realm of music and sound industries offered efficient channels for talks across cultural boundaries. The new media of sound recording and broadcasting enabled efficient communication between countries and was imbued with the political ambition of engineers working as diplomats with a shared consensus over music's power to foster peace and civilization.

Networks of acoustic engineers first became institutionalized through the creation of national associations, which were important for both military applications and the development of sound recording and broadcasting technologies.[30] In 1928, industrial scientists and engineers founded the Acoustical Society of America.[31] Its creation served as a model for European efforts to institutionalize the field and keep up with the growing power of American acoustic research. The following years saw the creation of similar bodies in several countries, including the Sound Foundation in the Netherlands in 1934,[32] and the Deutscher Akustische Ausschuß in Germany in 1936. And as more countries instigated national bodies to represent their acoustic interests, other states felt compelled to follow this example. In 1936, the general director of the French standards organization suggested to the minister of the posts that he appoint a national acoustical committee in order to put French acoustic science on an equal footing with its neighbors.[33]

The foundation of acoustic associations in the late 1920s and 1930s represented an effort to assert within the field of physics the legitimacy and utility of the research of the rapidly expanding community of acousticians. At a time when relativity theory and quantum physics attracted growing interest, acousticians started to feel increasingly marginalized due to the applied nature of sound knowledge. Young acousticians like Vern Oliver Knudsen and Harvey Fletcher confessed that they felt like "second-rate citizens" within the

American Physical Society, while older scholars, such as Dayton Clarence Miller, had a hard time coping with the latest developments in their field.

"The boundaries between acoustical science and commerce were hard to distinguish"[34] from the start, and the Acoustical Society of America secured large corporate support from the various branches of the sound industry, including AT&T as well as various musical instrument manufacturers, producers of architectural materials, and laboratories devoted to sound reproduction.[35] Similarly, the co-founders of the Dutch Foundation, Adriaan Daniël Fokker and Cornelis Zwikker, wanted to make acoustic research "more official, authoritative, and centralized," as well as to join forces with engineers in order to address pressing societal questions.[36] Among the 800 initial members of the ASA was an eclectic mix of acoustical engineers and physicists, psychologists, musicians, otologists, and phoneticians.

The activity of interwar acoustical societies focused on problems surrounding sound control, which called for the creation of standard procedures and units. The ASA's initial research programs included the development of sound-absorbing materials for buildings, noise abatement in cities and apartments, and the improvement of sound signal reception and reproduction for radios, phonographs, and telephones. Similarly, experts from the Dutch Sound Foundation were keen to provide solutions for a range of problems, including "the sound insulation of high-rise buildings, city noise, and the acoustics of concert halls and radio studios."[37] Reflecting the centrality of noise control in their program, the Foundation's official purpose included the design of official guidelines regarding home-building and construction material, and one of the society's first initiatives was the organization of anti-noise conferences.[38] Standardization was a key concern for all these programs. For example, the creation of a standard unit to record noise levels, the decibel, was at the center of interwar acoustic research, as was the manufacture of accurate audiometers.[39]

Although nationally organized and driven by international competition, acoustical societies were platforms for transnational exchanges, bringing together representatives of sound industries and helping forge powerful global networks. Annual meetings also provided further opportunities for interactions, as did the opening up of these associations to foreign membership. For instance, Erwin Meyer, one of Germany's leading figures in the field of acoustics, became a member of the Acoustics Society of America in 1931; and by the time of the 1939 London conference, although Meyer was invited to join the German delegation, he could not attend the meeting because he was otherwise engaged with the overseas body.[40] In addition to acousticians' travels, these societies started publishing journals that contributed to the internationalization of conversations on electroacoustics. Throughout the pages of publications like the *Journal of the Acoustical Society of America* and the *Akustische Zeitschrift*, reviews of foreign books and updates on the latest international research and apparatuses provided a diffusion of acoustic knowledge that transcended national boundaries.[41]

Late 1930s pitch standardization was also the result of an increased interest in sound among standards agencies. Standardization has been a part of the history of industrial capitalism since the early nineteenth century.[42] The creation of the French pitch was the product of France's broader efforts to unify weights and measures to facilitate the circulation of goods within its national territory and, later, on an international scale. Similarly, the 1885 Vienna conference was part of wider programs of international coordination: with this event, pitch joined a cohort of subjects negotiated within the late nineteenth century's "conference system," such as submarine cables, customs tariffs, or technical standards for railways.[43] By the turn of the twentieth century, national standards agencies started to appear in the United States and other countries. As organization studies' scholars JoAnne Yates and Craig Murphy have analyzed, standards-setting bodies "developed in response to the greater social complexity that accompanied the pressure toward the greater economic integration of industrial capitalism."[44] In 1901, several British engineering societies founded the Engineering Standards Committee. Overseeing cooperation among scientists, engineers, firms, and associations, it was the first private, voluntary national standardizing association. In 1918, the American Engineering Standards Committee emulated this model, cultivating a culture of voluntary adhesion and consensus as a way to establish American industrial standards.

With the development of telephone, radio, and mass broadcasting in the early twentieth century, standards agencies became urgently interested in acoustical standards. For example, the American Bureau of Standards initiated acoustic research in 1920, formally establishing the subject as one of its divisions in 1922. Three years later, in 1925, the institution opened a laboratory especially designed for testing acoustical materials,[45] which corresponded with representatives of the sound industries on a broad range of topics, including architectural acoustics, airplane noise, deafness, musical instrument making, and materials for sound absorption.[46] The American Standards Association also had an acoustical division, as did other standardizing bodies. As already mentioned, in France, the Association française de normalisation (AFNOR) suggested establishing the Comité national d'Acoustique, and by the end of the 1930s, the British Standards Institution also had an acoustic section.[47] In 1937, the ISA brought together members of these various national acoustic commissions to create Subcommittee $43/3^B$, and by the time Grützmacher launched the subcommittee's international consultation on pitch, individual countries could rely on their pre-existing acoustic networks to organize national negotiations on the subject. For instance, in May 1938, the British Standards Institution (BSI) held a meeting with representatives of broadcasting, instrument making, musical pedagogy, and physics, a prelude to the appointment of a pitch committee in charge of discussing concert pitch.[48]

In addition to providing national channels of negotiations, electroacoustics' new networks provided standardizers an abundant international arena in which to organize pitch conversations. In particular, although ISA Subcommittee $43/3^B$ was officially leading the negotiations, the broadcasters of the IBU played an important role behind the scenes. For over ten years, members of the IBU had been negotiating over radio wavelengths.[49] These discussions meant that, by the 1930s, broadcasters had a great interest in negotiations over pitch.

In December 1937 the IBU, in response to the 1937 international conference on acoustics in Paris, created a "mixed study group"[50] bringing together technicians and members of the IBU's Program department to study pitch standardization.[51] From the start, the IBU envisioned this study group as the unofficial leader of international discussions, whereas ISA Subcommittee $43/3^B$ would enforce the decisions reached by the study group. In the words of Hans-Joachim von Braunmühl, the director of the technical office at the German Reichs-Rundfunk-Gesellschaft (Reich Radio Society), one of Germany's representatives at the 1939 London conference, and a member of the IBU, when opening a back channel of negotiations with Hayes during the spring of 1938:

> The investigations carried out within the framework of [the IBU] to determine the standard pitch should serve to unify standard pitch for broadcasting for the present. It should further provide a contribution to the general international regulations on this question which, as you know, the International Congress has on its agenda. This congress, which works officially as ISA Sub-Committee 43 (International Standards Association), will naturally not achieve its results with the same speed as broadcasting people, who are used to working quickly. In spite of this, it might be of assistance towards the resolution of the question if the BBC could get into touch with the British

> Standards Committee, which acts as an official member of the ISA independently of the work done within the [IBU]. The German Committee of the ISA was appointed in Paris as the secretariat for electro-acoustic questions and the regulation of standard pitch.[52]

Along with Braillard, who officially represented the IBU at the 1939 London conference, several members of the organization attended the discussions as members of national delegations and on behalf of their various institutions. In addition to von Braunmühl, Balthasar van der Pol, the representative of Holland and a physicist at the Philips Corporation's laboratory since 1931, was an active member of the IBU's study group on pitch and provided ISA Subcommittee 43/3[B] with data as preparatory material for the London conference.[53] Finally, Hayes, although not a member of the British delegation, played an important role in the preparatory conversations organized by the BSI prior to the event.

The BSI pitch committee was well aware of the existence of this parallel channel of consultation among broadcasters, and used it as a tool to secure consensus in Britain. During one of the meetings of the body, the chairman asked engineers from the BBC to keep them updated on the IBU's proceedings. The two organizations also attempted to coordinate their efforts, as the BBC engineer Francis William Alexander reported after a meeting of the BSI pitch committee, noting that "the committee is very anxious to avoid any overlap between the two organisations."[54] Interestingly, however, although the BSI pitch committee frequently discussed the IBU's activities, these conversations of the meeting were never mentioned in the official reports produced by the BSI, perhaps in order to avoid giving the impression that the two organizations had engaged in a concerted effort to give the broadcasters more of a say in the conversations.

The covert pitch diplomacy of these broadcaster networks showed a surprising resilience, and communications on the subject continued during the early months of the war, even between sound engineers of belligerent nations. In February 1940 Hayes, following up on a query from the BSI, asked Braillard for information regarding the enforcement of the agreement of May 1939. Braillard responded quickly, detailing a conversation he had conducted with von Braunmühl, in which the German had reported that "he had equipped studios with a central distribution of the pitch 440 broadcasted through loudspeakers and that he had, in addition, made portable apparatuses."[55]

That this commitment to sound standardization proved to be so robust, even in the face of war, demonstrates how the relationships between these engineers were not merely professional, but had developed into friendships. After a meeting of the IBU in Montreux, in April 1939, Braillard summed up this sense of camaraderie in a note to Hayes, reflecting that "This conference may have been difficult, but it was nevertheless a 'good time' spent among good friends."[56] Through difficult times and international conflicts, these personal connections kept conversations going between Germany and Britain.

Pitch diplomacy extended beyond Europe. Among the decisive conversations that took place between 1937 and 1939 was a secret bilateral agreement between Britain and the United States. Along with other questions, the BSI's approach to pitch standardization was part of a broader effort to build relations with the United States, and the two countries' rapprochement amid rising international tensions. In June 1938 Charles Le Maistre, the head of the BSI, had advised Britain to cultivate a shared understanding with the United States before responding to Grützmacher's initial call for recommendations.[57] Over the following months the BSI, following this advice, labored to secure Britain's adhesion to *A* 440 and to collect detailed information about the United States' approach to the standardization process.[58] In October 1938, as part of these efforts, the BSI organized a meeting between the members of its pitch committee and one of America's leading acousticians, Vern Oliver Knudsen, a co-founder of the Acoustical Society of America.[59] It is unclear whether Knudsen visited London to attend this meeting, but the exchange proved influential, as following it Britain and the United States submitted a joint proposition to ISA Subcommittee $43/3^B$ through their respective bodies: the BSI and the Acoustical Society of America. Revealingly, this shared position was not acknowledged in ISA Subcommittee $43/3^B$'s official documentation. Since Germany was eager to appear the leader in pitch conversations, the ISA's German Secretariat did not want to recognize the Anglo-American consensus.[60]

SOUND GOES LIVE: RADIO AND THE MAKING OF CONCERT PITCH

The long-awaited agreement between countries, including some like Britain and the United States which had never before participated in international pitch negotiations, was not the London conference's most important outcome. At several junctures before 1939—Vienna in 1885, or the United States and Britain's efforts to adopt the French pitch—increased international regulations had seemed at hand. Infallibly, however, such moments had provoked controversy over the measure to be adopted or whether pitch should be defined as an absolute or a relative value indexed on specific temperature levels. These questions were still up for debate in London in 1939. Preparatory studies for the conference and conversations held during the meeting, however, helped fashion a growing international consensus over measurement methods and units. In other words, sound engineers not only agreed on the choice of a certain pitch, but more fundamentally harmonized their views about the phenomenon itself. Their exchanges resulted in a complete redefinition of pitch as a scientific object, one shaped and defined by the new media of sound recording and broadcasting.

The production of this new definition of pitch represented a little epistemological revolution. Crucially, instead of focusing as earlier acousticians had on the "initial frequency" of musical instruments or tuning forks outside any musical context, pitch studies of the late 1930s placed pitch within the frame of actual musical performance. The new regime of recording enabled the analysis of pitch over the course of an entire work of music, and radio afforded acousticians unprecedented access to tuning practices across the technologized world. Grützmacher's second letter on behalf of ISA's subcommittee $43/3^B$, circulated on 29 December 1937, emphasized this reconfiguration of pitch as a scientific object, laying the foundation for some two years of experimental investigations across Europe. In this document, Grützmacher, as head of ISA Subcommittee $43/3^B$ and director of the Physikalisch-Technische-Reichsanstalt's (PTR) acoustic laboratory, presented the results of a study that his colleague, Werner Lottermoser (1909–1997), had directed and published in the *Akustische Zeitschrift* just a few months before.[61] As Lottermoser explained:

> In order to ascertain whether the European orchestras keep the concert pitch a [*A*] 435 Hz. vibrations per sec., as agreed at the International Conference for Concert Pitch in Vienna, tests were made to check the musical broadcasts of as many stations as possible with a view to measuring their concert pitch.[62]

The PTR's involvement with pitch was not new. After the Vienna conference of 1885, the institution became Germany's control center for tuning forks.[63] Lottermoser's study, however, with its emphasis on sound recording and broadcasting, was part of the Third Reich's new scientific agenda. The experiment took place in the recently opened acoustical laboratory. Created in 1934, this institution was very much a response to the Nazi state's demand that the nation's physicists abandon pure research in favor of practical experiments that would support domestic interests.[64] In this way, the PTR's scientific program became increasingly politicized. The work of the acoustic laboratory was primarily aimed at undergirding Germany's rapid militarization and war preparation, contributing to the production of acoustic mines and acoustically controlled torpedoes.[65] At the same time, it was conceived of as an aid to the development of Germany's national music and sound industries, including instrument making, broadcasting, recording, and motion pictures, itself useful to creating Nazi propaganda.[66]

Grützmacher was appointed head of the institution's acoustic laboratory amid the reconfiguration of the PTR's scientific agenda. An authority on acoustics, he was not only experienced in scientific research, but had worked in commerce and industry. This straddling of industry and university research was typical for acousticians of Grützmacher's generation. A doctor in physics, Grützmacher started working on acoustics while he was an employee at the Telegraphentechnische Reichsamt.[67] In 1930 he became the director of the acoustic laboratory of the Reichspostzentralamt, where he elaborated a new practice for sound analysis and, in collaboration with Erwin Meyer, a new arrangement to measure the spectrum of musical instruments.[68]

Grützmacher pursued these sonorous experiments at the PTR, where he developed an apparatus for measuring and visualizing the fundamental pitches of speech that was to become a reference for phonetic research after World War II.[69] Starting in 1935, Grützmacher also initiated experiments on the pitch of musical instruments, including pianos, organs, and wind instruments.[70] In these endeavors he collaborated with Lottermoser who, having just received his doctorate in physics with a thesis on the sound of reed pipes, was keen to volunteer at the PTR. An enthusiastic admirer of German organ building since his youth in Dresden, Lottermoser embraced the career of a physicist to try to explain the fascination that organ sounds produced on him, and indeed devoted most of his career to the study of the instrument.[71] Grützmacher valued Lottermoser's training as a musician, which provided the laboratory with essential skills for experimenting on musical instruments.[72] Lottermoser's profile and particular interest in music and aesthetics in turn shaped the PTR's approach to acoustics and subsequent intervention into the question of pitch conversations.

The PTR's work on the pitch of musical instruments was very much part of the institution's new nationalistic scientific policy. Beyond its martial applications, the institution's acoustic research also furthered cultural and commercial endeavors. In 1935, following a set of experiments on the tuning of grand pianos, Lottermoser suggested the adoption of a new way of tuning instruments. Having established that equal temperament is neither musically pleasing nor actually achieved in practice, he promoted a new tuning system based on the relations between pure fifths.[73] Lottermoser's aim was to help improve German instrument making, as a subsequent experiment on the tuning of organ pipes made clear. Drawing on the measurement of ancient and modern organ pipes, Lottermoser emphasized the influence of materials on their timbre, itself due to discrete pitch variations. Builders of the past, he argued, had made good use of lead or lead alloy, which gave organs "a slightly vibrating, round tone. . . . Hearing," he continued, "needs a certain tremor in order to perceive a sound as musically beautiful . . . : *Cum grano salis*."[74] With this research, Lottermoser not only highlighted the previously overlooked relation between the material, pitch, and tone color of organ pipes but hoped to contribute to Germany's industrial and cultural dominance by improving the quality of the nation's instruments.[75]

The apparatus developed by Lottermoser and Grützmacher over the course of PTR's research on musical instruments allowed for a new kind of observation of pitch as a scientific object. Earlier studies of pitch that had relied on beat counting to record frequency were time consuming and had to be repeated many times to ensure accuracy. It was already immensely difficult to measure the pitch of a tuning fork or of a musical instrument in a laboratory, outside of any musical context—let alone to capture the sounds of a live performance featuring multiple instruments or notes at once, as was the case in most concerts. To overcome the challenge of sound's ephemerality and to record musical pitch within the context of musical performances, Lottermoser used as a kind of pitch recorder a valve oscillator connected with an amplifier and with a device that measured frequency. By modulating the valve oscillator's frequency to match the pitch of certain tones, "which distinguished [themselves] by frequent occurrence, for instance in final chords, organ-points, etc.," Lottermoser was able to adjust to the fast pace of live music and perform indirect measurements in situ.[76] In this setting, recording provided acousticians with both the musical material to measure, and the means to do so.

Although it relied on the ear and actions of the experimentalist to match the pitch of the variable tone connected to the measuring bridge with that of the musical performance, Lottermoser's apparatus was reminiscent of the techniques developed in the field of physiology and medicine in the last decades of the nineteenth century, by which scientists measured dynamic physiological phenomena through the use of self-recording instruments and graphic methods.[77] In 1847, the physiologist Carl Friedrich Wilhelm Ludwig created the kymograph, an apparatus that recorded the modification of arterial pulse with a stylus on a rotating drum; and in the following decades, several researchers emulated this technique to measure breath, brain activity, speech, and other phenomena unfolding over time. Interestingly, music had served as a model for these endeavors. In their ability to record "heterogeneous but simultaneous processes in relation to one another," graphical representations reminded physiologists of musical scores, "in which the different parts of the orchestra and the choir are represented synchronously."[78] Like these systems, which brought the movement of life into the purview of metrology, Lottermoser's method had the ability to capture music as a "live" phenomenon.

To be sure, diverse methods had been used to "record" sound since the middle of the nineteenth century, the first one being the Lissajous method of visualization, presented in chapter 1. But Lissajous's curves only enabled the measurement of a tuning fork producing a single pitch. In the last decades of the nineteenth century, electricity had provided acousticians with the means to maintain tuning forks in vibration for indefinite stretches of time, thus allowing for longer, far more accurate, measurements than before. But recording pitch had remained an activity entirely separate from musical performance. Through the use of recording devices both as a way to capture music and bring it into the lab, and as a means to measure these displaced sounds, interwar studies on musical pitch introduced a radical new perspective to the debates. Pitch was recast as a "live" phenomenon and joined the pulse, the breath, and the brain as phenomena that articulated time and space on a far broader scale than ever before (see fig. 4.1).

Although Lottermoser's study differed from past experiments in its ability to scrutinize ephemeral sounds thanks to the mediation of electroacoustic technologies, the production of data ultimately raised the same questions that Ellis, Cross, and other acousticians had confronted. To make sense of the different frequency values observed, Lottermoser had to determine their relations from a musical point of view: he had to select a tuning system, whether the tempered scale, just intonation, or meantone temperament. Neither the seemingly more "realistic" conditions of observation that broadcasting afforded, nor the accuracy in measurement enabled by his electroacoustical apparatus, could solve this epistemological puzzle. On the contrary, the conversion of data on "live" performances required an even greater level of theoretical abstraction than the measurement of "initial frequency" hitherto considered by acousticians. Whereas earlier acousticians had dealt

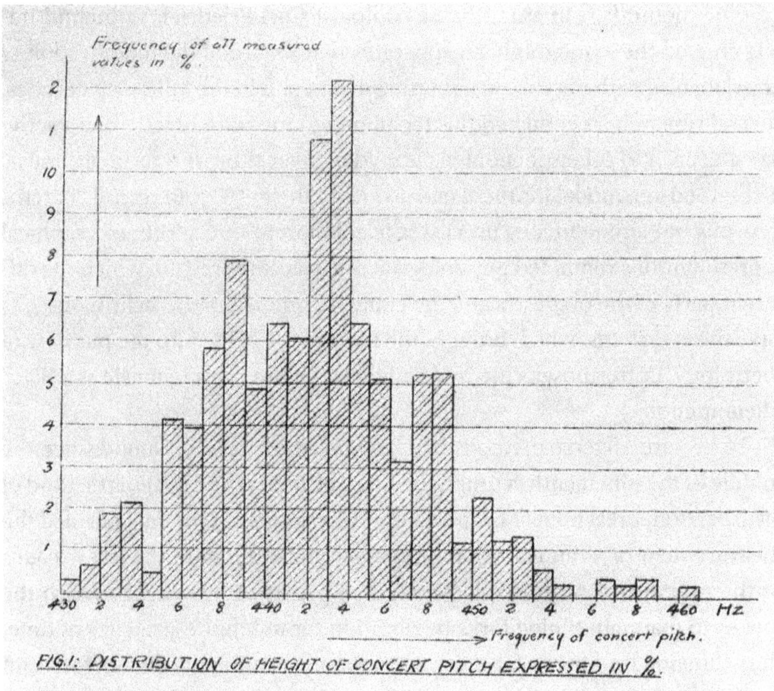

FIGURE 4.1. "Distribution of height of concert pitch expressed in %." Extract from letter from the German secretariat of ISA subcommittee 3. Electroacoustics, Musical acoustics of ISA 43. Acoustics and Report on the measurement of the concert pitch a' in musical broadcasts, submitted by Werner Lottermoser, Phys.-Techn. Reichsanstalt, Berlin. Courtesy Surrey History Center.

with only a few standards (essentially low *A*, middle *A*, low *C*, and middle *C*), Lottermoser wanted to analyze all recurring tones. This meant the tonic, dominant, and subdominant—in other words, any note with a structuring role within a given musical work. In doing so, he expanded the place of temperament in the production of pitch data, since now he had to account for the relationship between all twelve tones and *A*.

To meet this epistemological challenge, Lottermoser came up with a fiction proportional to the alleged objectivity of his observations. To measure pitch in the context of musical performances, Lottermoser first produced a systematic taxonomy of tuning and applied it to the entire range of instruments considered in his study, reasoning that "In order to be able to make statements as to the height of the tone of the concert pitch, tempered intervals were taken into account in the case of piano and organ recitals, and pure intervals from the measured tone up to a' were taken into account for chamber music (without piano) and chorus."[79] In other words, to compare the results he obtained from measuring diverse tones of the scale, Lottermoser embedded them within specific tuning systems according to their instrumentation and genres. This meant that he applied different proportions in his calculations for different musical works: for example, he used the ratios for the tempered scale when considering an organ performance, while he employed just intonation when analyzing a string quartet concert. Lottermoser, however, provided no evidence that the tuning systems he selected were the ones actually in use. In fact, as he knew all too well himself, both tuning systems were mere theoretical ideals, never fully attained in practice. In short: by expanding the territory of his measures through a seemingly more objective approach, Lottermoser moved even further from the reality of musical performance, reaching a new level of abstraction.

Lottermoser's study was not unprecedented.[80] But thanks to the publicity it attracted through ISA documents, the experiment was influential in transnational conversations on pitch. In the months following, Grützmacher sent a letter containing a reprint of Lottermoser's study to European standards agencies, and the physicist's approach became the standard way of investigating pitch. Equipped with frequency measurement apparatuses for controlling radio waves, as well as research departments conducting acoustic experiments, radio stations were ideally suited to lead in the research on broadcast musical performances. The BBC had the frequency of its various studios' organs measured at its receiving station in Tatsfield.[81] Likewise, the Reichs-Rundfunk-Gesellschaft made similar measurements of its own orchestras.[82]

When the IBU's pitch study group commenced work on the question of musical standardization, it requested its members to take measurements and send information about their country's tuning practices. The institution's Checking Center, established in 1927 in Brussels in order to solve problems of radio wave interference and cross-border transmission, was a leading venue for frequency regulation in Europe. Throughout the fall of 1938, this center hosted a large-scale study into musical pitch similar to the one conducted at PTR the previous year. Balthasar van der Pol, head of the Philips acoustic laboratory, led over a hundred observations of musical broadcasting. The measurement protocol is evidence of the physicist's attempt to secure increased mechanical objectivity along the same lines as Lottermoser's study. This time, instead of the ear, it was the eye that compared the pitch of recordings with that used as a standard. As with the physiologists' experimental systems, van der Pol tested the pitches of European orchestras by visualizing both the examined recording and the frequency of a measuring device. His method consisted of passing that portion of a recording's frequency wave that fell between 400 and 470 hertz through a cathodic oscillograph. Whenever a pitch corresponding to this frequency range was played, a luminous point would then appear on the screen of the oscillograph. This pitch could be compared with the one produced by a measuring device: a string electrically maintained in vibration as well as visualized on the oscillograph's screen.[83] Besides replacing the ear with the eye, van der Pol's study allegedly promised greater accuracy than Lottermoser's in the measurement of certain instruments, notably organs (the accuracy of van der Pol's method was of 0.2 cycles per second for organs, versus 0.5 for Lottermoser's).[84]

André Labrousse, engineer at the Technical Department of the French Post Office, led a comparable experiment on behalf of the French Acoustical Committee.[85] While also emulating Lottermoser, Labrousse, like van der Pol, replaced the German physicist's aural technique with visual observations. Asserting that one should not "rely on subjective appreciation," and that "it is evidently preferable to have available recorded data which can be studied at leisure," the French study relied on the visualization of pitch.[86] Labrousse, however, lacked the German physicist's musical training and employed others to listen for him and indicate on a musical score which notes (all middle *A*s) he should measure.[87]

Differing in methods and apparatus, the data produced through these studies was consistent on one point: the pitches in use across Europe were undoubtedly above 435 hertz. And although initially sound engineers were primarily concerned with pitch's geographical inconsistency, they became increasingly aware of, and sympathetic to, their predecessors' concerns over the threat that this pitch escalation presented to Western musical heritage. To some extent, these anxieties reflected the solidification of musical canons and the connection between new technologies of sound recording and broadcasting on the one hand, and the cultural celebration of historical masterpieces on the other. Yet this disquiet also shows how, for all that technology radically transformed in the 1930s, important continuities remained between interwar efforts at musical standardization and those made before 1914.

Lottermoser, van der Pol, and Labrousse's investigations not only converged on how they defined pitch. They also produced similar cultural definitions of this physical phenomenon, based on the technological opportunities that European broadcasting infrastructure provided along with a shared appreciation for Western classical music—as opposed to jazz, dance music, or other kinds of musical practices that were represented on the radio.

All three displayed similar aesthetic limits, focusing on classical orchestral and organ performances. Indeed, van der Pol's experimental data for the IBU referred to specific musical works and specific categories of instruments, including harpsichords. The two examples of detailed "analysis" it presents were performed at the Checking Center. The first is the Dutch conductor Willem Mendelberg's performance of Beethoven's first piano concerto on 27 October 1938 (probably a recording of the Concertgebouw orchestra). The second one examines Birmingham's municipal orchestra's concert of 27 October 1938 featuring a "Bach concerto for oboe and orchestra."[88] At a time when broadcast schedules included significant segments of light and popular musical repertoires as well, this focus on historical masters shows that, despite the use of new media in pitch negotiations, reformers remained predominantly interested in pitch within the canon of classical music.

In other words, these studies shaped not only scientific but also sociopolitical and cultural definitions of pitch. Of course, one can only speculate about what would have happened if jazz and dance music had informed the production of interwar data on musical pitch. Would the result of acousticians' measures have much differed? Would standardizers have come to a different conclusion in 1939 and, if so, which? Such questions will remain forever unanswered, but it is nonetheless important to keep in mind that as debates were becoming more and more connected with contemporary musical practice, standardizers continued to cultivate a privileged relationship with their cultural past and classical masters. Contrary to the word's etymology, data is never just out there: rather, it is already a construction that carries within it the prejudice of those attempting to gather the information.[89]

In addition to reflecting the cultural bias of acousticians in charge of tracking pitch during concerts, measurements conducted by them were initially constrained by the limitations of the 1930s broadcasting apparatus and contingent on the state of European technical integration. In the opening lines of his article, for example, Lottermoser defined the parameters of his study, acknowledging that "As it was considered important to have a regular, strong reception, the stations at a greater distance and the weaker senders had to be left out of consideration for the time being." His article subsequently focused on Belgium, Germany, Denmark, Britain, France, Holland, Italy, Yugoslavia, Norway, Austria, Poland, Switzerland, Sweden, Czechoslovakia, and Hungary. Van der Pol's investigation was geographically more limited, considering only Dutch, British, and German programs. Finally, Labrousse measured performances by French, English, German, Dutch, Swiss, and Italian orchestras.[90] Just as nineteenth-century and early twentieth-century pitch investigations had been dependent on the ability of investigators to travel,

FIGURE 4.2. "Reception at Lyne, Capel, Surrey, the home of Captain Evelyn H. T. Broadwood, M. C., M. I. M. T., to delegates of the international committee on the standardisation of concert pitch, invited to England by the British Standard Institution (ten countries are represented on the committee). 587492. The Tudor Singers' Recital. S&G. 14/5/39. ETF" (legend on the reverse of the photograph). Courtesy Surrey History Center and Alpha Press.

pitch scholars from the late 1930s had to rely on existing radio infrastructures, themselves the result of previous diplomatic encounters, to secure information on international musical practices.

Nowhere was this historicist cultural embedding of concert pitch more apparent than at the 1939 London conference. On 13 May, Evelyn Broadwood, as one of Britain's delegates, received the representatives of all the other nations at his country house in Surrey. The highlight of this party was immortalized in photographs that show the delegates listening attentively to the sounds of mostly English songs from the Renaissance performed a cappella by the Tudor Singers (see fig. 4.2).[91] Whether or not the singers were performing to the brand-new international pitch, this depiction of international harmony reflects the agreement of nations on a fundamental point: the value of early European music.

Through the creation of a concert pitch, broadcasters defined a musical world based on technical developments, diplomatic relations, and a specific cultural conception of musical canons. This norm was nevertheless meant to determine the tuning of the entire world, confirming, once again, the Eurocentric and hegemonic ambitions of standardizers. A year after the London agreement, the secretary of the BSI proudly reported to the BBC that "the South African Broadcasting Corporation has not only adopted standard pitch, but is giving it by the generator."[92] The 1939 resolution appeared to be having its intended influence.

ENFORCING STANDARD PITCH: MUSICAL PRACTICES UNDER CONTROL

Radio was not only the precondition for pitch standardization in the late 1930s. It also offered a new means to enforce the standard. The resolutions that Subcommittee 43/3^{B} adopted consisted of two parts: the new standard, and a more elaborate appendix comprising six questions that formed a "programme of study regarding the methods to be adopted to ensure the practical observance of the standard concert pitch."[93] This appendix was intended to be used as a foundation for groups implementing the standard across its various contexts of application.

Drawing on the new data produced through the pitch studies of the 1930s, the questions demonstrated an increased appreciation of how musical instruments behaved over the course of a concert. For example, they sought to resolve how the impact of temperature on different categories of instruments could be controlled, as well as the kind of tuning devices that orchestras should use, and the procedures that would best regulate the tuning of instruments and tuning forks on a national and international scale. Some of the answers to these questions indicated the increased authority of scientific knowledge over musical practices. The IBU pitch study group wrote a response to the questionnaire on 5 June 1939. First, they urged a change in the routine of musicians such that during musical performances wind instruments be positioned so as to conform with the variation of a room's temperature and thus maintain a constant pitch throughout a concert. They also suggested minor refinements of the standard as a way of responding to specific instruments' behavior: namely, that wind instruments be tuned "1 to 2 c/s above the tuning note." Finally, the study group challenged the traditional role of the oboe in setting an orchestra's *A*, recommending instead the broadcast of the standard through "radiophonic transmitting stations."[94]

If these remained merely aspirational suggestions, at least for the time being, a few studios took the initiative of developing musical practices under just such scientific control. Meanwhile, radio broadcasting began implementing the standard by taking over the function that tuning forks had hitherto filled and manufacturing a new kind of prototype: radio waves. This manifestation of concert pitch through broadcasting would have consequences for both the world of music and the public's experience of technology. Tuning, in this context, was a reciprocal process.[95]

Already on 22 December 1937, in a memo to various services of the BBC, Hayes had proposed a setting adopted in Vienna that enabled the broadcast of "a central frequency standard in the Control Room, a note that could be produced on the loudspeaker of any music studio, and that our orchestras should tune to."[96] The conductor Adrian Boult received this idea with great enthusiasm.[97] A few weeks before the 1939 conference in London, the Swiss broadcaster Maurice Rambert wrote to Balthasar van der Pol, explaining how he was personally in favor of having radio play a central role in the process of pitch standardization:

> I asked that . . . when our special clock for the hourly signals of Swiss broadcasting was to be built that these signals be given on the frequency of 440 Hertz, or on the "A" of the tuning fork, so that they could be useful not only to those who want to adjust their watch but also to musicians and piano tuners . . . [This] idea [is] probably more important than it seems since it is about having broadcasting get everyone to agree by setting the tone for the Concert of Europe.[98]

As Graves summarized in his opening speech in 1939,

> it is a problem for the musician to decide on the standard pitch which he finds most convenient. It is then that the aid of the technician must be sought in order that the musician may be helped to use and maintain the pitch he has chosen. I think that the present departure from the Vienna agreement of 1885 may be due in large measure to the fact that the assistance of technicians was not at the time available and has not been used since to maintain the agreed figure.[99]

After the 1939 London conference, the BBC began working more actively to introduce the new standard in its studios, a step Germany had already taken by providing musicians within its premier studios with broadcasts of the standard and portable electronic tuning devices calibrated to it.[100]

The BSI had been one of the most vocal advocates for the use of radio as a means to enforce concert pitch. At the end of the British-American joint proposition sent to ISA Subcommittee 43/3^{B}, two notes detail the technical side of the countries' position. The first concerns temperatures and various categories of instruments, and the second specified that "Broadcasting authorities should be invited to broadcast the international standard of musical pitch at convenient intervals during their programs."[101]

In his report of the meeting during which the text was written, in October 1938, Alexander explained that this plan had its origins in the United States, in discussion with the American acoustician Knudsen.[102] He then assured his colleagues that the musical world or, rather, the music trade industry, was enthusiastic about this project, informing them that

> Practically all the Committee present, especially those representing the Music Trades, thought this was an excellent idea from their viewpoints, and that the same should be done in this country. I was asked if this were possible—I guardedly said that probably something could be arranged, to have the standard pitch broadcast, not of course every ten minutes, but perhaps occasionally during the course of a week.[103]

This idea (which would only materialize after the war, in 1946[104]), went a step further than the solution adopted by the United States. The station from which the American Bureau of Standards emitted its 440 hertz signal was strictly dedicated to the broadcasting of standards, and was chiefly designed for instrument makers, for professional purposes. In other words, pitch was not part of the daily landscape of radio listeners. The BBC, on the other hand, suggested making *A* 440 part of the general public experience of radio. Of course, it was not the first norm to be broadcast through radio: the medium had long diffused standards of time known as the "Greenwich pips," a series of six short tones broadcast at one-second intervals on several of the networks' stations, introduced by the BBC in 1924.

In order to provide musicians and instrument makers with the technology broadcasters used to tune radio waves, Alexander suggested aligning the BBC's standard frequency signals with concert pitch, replacing the existing signal of 1,000 hertz, broadcast every morning before the commencement of programs with *A* 440. In this way, he argued, the signal would unite musical practices throughout the British Empire.[105] In his memoirs, BBC engineer Edward Pawley, who was working for the Overseas Engineering and Information Department at the time, observed that his company's decision to adopt 440 was because it would be easier to broadcast than the country's own version of the diapason normal, *A* 439, which had been the official pitch in Britain since 1895—though there is no archival evidence to support Pawley's claim.[106]

Documents related to the BBC's efforts to produce a reference signal provide interesting insights on the various questions and operations implicated in the making of such a standard. Alexander's colleagues accepted his suggestion, and in the winter of 1938 the BBC started working toward the implementation of this decision. In addition to questions regarding the level of accuracy of the signal,[107] engineers considered which sound would be best suited to tune the musical world, reaching a solution through conversations with the music department.

One of the questions debated by engineers and musicians concerned the nature of the signal. While working on the subject for the Reichs-Rundfunk, von Braunmühl had suggested that to suit musicians' ears, "the note should not be sinusoidal and they had found that 50 per cent harmonics were desirable."[108] After discussing the issue with the BBC Music Department, however, Alexander established that "the broadcast of the pure note [was] desirable for the following reasons: trade people have been accustomed to using tuning forks for a very long time which emit a relatively pure note. The broadcasting of a complex note would be, [Music Department Deputy Director Thatcher] thinks, confusing to them."[109] Where von Braunmühl conceived of his *A* signals as rendering something of the complex sounds of musical instruments, the BBC modeled its standard on traditional technologies such as tuning forks. The intended audiences of this radio signal were not musicians but rather manufacturers: "The question of the desirable type of note for tuning orchestras," Alexander concluded, "is a different matter."[110]

A final consideration related to the format and exact modalities under which the tone should be emitted. In this regard, musical considerations were not the only concerns broadcasters had on their minds. Responding to a suggestion Rambert had made to the IBU in April 1939, advocating the use of time signals to broadcast concert pitch, van der Pol suggested that these were too short to enable a listener to tune an instrument.[111] In Germany, the Reichs-Rundfunk diffused the standard for five minutes. But BBC engineers were not sure how to introduce this signal without causing too much disruption in their existing schedule. Although the main beneficiaries of this new signal, musicians and makers would not be the only ones to hear it, and some British engineers considered it potentially inconvenient for their listeners.

Throughout the summer of 1939, sound engineers from the BBC prepared for the broadcasting of a signal of 440 hertz from their long-wave station at Droitwich. After consulting with the nation's professional organizations, however, they decided to avoid rushing a decision that engaged a variety of fields beyond those of broadcasting. The outbreak of the war then curtailed their activities, delaying their project until 1947.[112] Starting the first of February 1939, while British engineers were still discussing the production and reception challenges involved, the Deutschlandsender based in Zeesen issued a daily transmission of the normal pitch—the same signal being broadcast in the United States.[113]

This dissemination of the standard demonstrates how the new episteme of pitch did not remain a purely theoretical construction. Unlike past efforts at musical integration, the negotiations of the late 1930s seriously addressed the question of implementation. Although interrupted by the war, the sound engineers' agenda to enforce pitch had a lasting impact on musicians, listeners, and instrument makers. As concert pitch was manufactured through the world of electroacoustics, concerts became places increasingly under the regulation of science.

Pitch standardization was part of the global cultural and political agenda of broadcasters in the 1930s and provided engineers with the opportunity to demonstrate how they could contribute to the maintenance of international peace. Even after war broke out in September 1939, several technicians asked to be allowed to continue their investigations. In 1940, after the BBC mobilized for military purposes and called a halt to all work on pitch standardization, Hayes refused to abandon completely his efforts to construct tone generators that would assist the BBC's orchestra in tuning to the new standard. In February 1940, he wrote to one of his colleagues that he was "glad to think that [the new British Standard Concert Pitch] is not dead, and I am afraid I cannot sympathize with your wish it should die."[114]

CONCLUSION

For the last ten years, conspiracy theories have blossomed over the internet, denouncing the harmful character of concert pitch *A* 440 and its alleged connections with Nazi Germany. Blending historical facts—namely, the loaded date of the standard's adoption, 1939—with New Age considerations about sound and well-being, several bloggers claim that the creation of concert pitch was an exercise of sonic manipulation, one by which the Nazis took control of people's emotions, and that, as a result, "the Western psyche has been antagonized by Nazi mood depressants for 75 years and counting."[115]

While previous works have unraveled the connections between 1930s sound technologies, German soundscapes, and Nazi politics, the history of pitch standardization is more complex than these bloggers and their forerunners suggest.[116]

As this chapter has shown, the engineering of an interwar musical standard was the outcome of complex processes of political integration as much as it was the product of German nationalism. If Germany, as secretary of ISA's acoustic section, was indeed in charge of organizing pitch negotiations and the 1939 conference in London, the agreement reached at this meeting owed much to broadcasters' decade-long efforts to integrate Europe through the development of telecommunication infrastructures. Furthermore, acoustic bodies and standards agencies officially in charge of the talks were not the only ones involved in pitch conversations. The IBU especially created additional connections between European sound engineers and an unofficial platform for conversations that sped up the negotiation process. Due to the involvement of this new category of actors in the late 1930s pitch negotiations, radio played multiple roles, being at once an imaginary space built on political and aesthetic premises of cultural integration and homogeneous listening experiences across borders, a network enabling one of the most striking diplomatic achievements in the modern era, a source of information about a transnational musical space, and a tool for further integrating European sonic practices.

If one had to determine the birthplace of concert pitch *A* 440, the United States would be a more obvious answer than Germany. Their intervention was decisive less for the choice of this specific frequency than because the country's decision to broadcast a signal to implement the new standard set the tone of subsequent European conversations. After the London conference, radio was used as a tool to enforce the standard in Europe just as in the United States. Reflecting a difference between radio developments in Europe and the United States, national European broadcasters went beyond the American initiative, broadcasting 440 hertz signals on the main national radio programs as well as within radio orchestras—and thus in musical broadcasts. While most of the activities examined in this chapter were abruptly interrupted by the outbreak of the Second World War, nations continued their efforts even during the war, and in 1945, worldwide efforts to implement pitch standardization started again with particular intensity.

CHAPTER FIVE

Postwar Aftermath

Confirming an Embattled Standard

> What do the violins and the wind instruments do and feel when tuning to A=440 c.p.s.?
>
> LLOYD S. LLOYD

What remained of efforts to standardize pitch after the devastating experience of World War II? How did the world's new divides shape the making of the standard? In answering these questions, this chapter sheds light on the final stage of pitch negotiations. Beyond politics, aesthetics complicated pitch conversations, confirming the unique character of the standard in comparison with other international norms—and, thus, the impossibility of submitting music to the rules of techno-science.

On 18 October 1945, the *New York Herald Tribune* reported in an article entitled "French Musicians Call for Stabilization of A" that France's Direction of Arts and Letters, the country's governmental office in charge of cultural matters, planned to organize an international conference to solve the problem of pitch escalation. Six years had passed since the agreement had been reached in London on concert pitch *A* 440, and significant strides had been made during and after the Second World War to create a more permanent structure to coordinate standardization worldwide. The lack of standards for screw heads, for example, had engendered considerable costs during the conflict. On an American initiative, the United Nations Standards Committee was established in 1944. The International Organization for Standardization, or ISO, the new federation overseeing standards' international coordination, would emerge in October 1946,[1] with the British Standards Institution as its acoustic secretariat. Meanwhile, the BSI's International Press Bureau brought the *New York Herald Tribune*'s article to the attention of Percy Good, the organization's director, and he in turn responded to the news by re-establishing contact with the various institutions that had participated in prewar negotiations. Good hoped to take the lead in international conversations on musical standardization and make sure that France did not ruin the labor that the BSI and its collaborators had invested in creating the standard before the war. In the letters Good sent to his collaborators in 1945, he indicated that the BSI was eager to organize a new international conference in the near future to validate the 1939 decision. As he wrote to the BBC engineer Edward Pawley, one of Britain's main artisans of the 1939 agreement: "I think all the people in this country who took such an interest in this work at the time would be sorry if it were allowed to become inoperative through a lack of a little effort on the part of those immediately responsible."[2] Securing the conservation of the 1939 agreement, as it turned out, required far more than just "a little effort."

"CAN WE HAVE AN 'A'?"

ISO's ability to set the tone in international pitch negotiations did not grant the organization the power to implement *A* 440, as standard-setting bodies increasingly realized after 1945. Although cultural authorities in France were threatening a revision of the 1939 decision, standard-setting bodies such as the BSI continued their efforts from the late 1930s to enforce the new measure across a broad range of musical contexts, including music studios, musical instrument workshops, schools, movie theaters, dance halls, and opera houses, as well as within the wide and heterogeneous assortment of objects and subjects that generated or perceived sound, from tuning devices to musical instruments to listeners' ears. As standardizers recognized after the war, this task required far more than the design of 440 frequency signals and other electronic tuning devices. Along with the production of these technologies, the implementation of the 1939 resolution demanded the active participation of the full range of musical parties.

The need to coordinate broad portions of society resulted in the cultivation of practices that set concert pitch apart from most units of measure. The success of other technical and industrial standards often depended on their invisibility, their ability to imperceptibly reshape scientific, industrial, and social practices. Pitch standardizers, on the other hand, hoped to impose the sound of *A* 440 across society. The BSI tried to use radio as a tool for technical purposes, but also for social engineering. In the hands of British reformers, the BBC became an instrument of propaganda intended to convince musicians to give life to *A* 440 through their performance practice.

On 11 December 1945, two months after publication of the *Herald Tribune* article, Ethel Wiggins, the secretary of the BSI's acoustic committee, wrote to Edward Pawley, the Corporation's head of engineering secretariat, to inquire about the steps that the BBC had taken to enforce the 1939 agreement and ask if there was "any possibility that the full recommendations . . . be carried out in the near future."[3] Wiggins further explained that although the BSI was eager to take the lead in the negotiations and would soon attempt to organize a second international conference on musical pitch, its members believed that since the ISO, the new federation overseeing international standards, had only been established three months earlier, the time was not yet ripe.

Furthermore, the BSI wanted to collect further information about various nations' efforts to introduce the 1939 standard. Not only were countries slow in responding to the BSI's 1945 inquiry, but amid postwar reconstruction, standardizers faced considerable economic and technical difficulties. Pawley's reply to Wiggins's letter pointed out that while the corporation's main orchestra had been playing at the selected frequency for some time, the BBC had not been able to proceed with the other points of the program that its members had established before the war.[4] Despite his and his colleagues' eagerness to continue their work from the late 1930s, Pawley warned the BSI that the BBC's "reduced Engineering Staff is very fully occupied on major problems of great urgency."[5] A few weeks later, he reported that the assistant controller engineer had ruled that the Corporation "should not undertake" the work of implementing standard pitch *A* 440 "until we have cleared some of the 'vast programme of designs work involved in putting out programmes,'" adding that if the BSI's request were to "be tackled," it should be "as a by-product of another job."[6] The BBC only started broadcasting *A* 440 signals in May 1947, and at this time only one of the Corporation's studios emitted this sound.

Technical and economic problems were not a struggle only in countries most affected by the war. It took years for Australia to introduce a standardized concert pitch, despite the BSI's influence over the nation's standards association and the Australian Broadcasting Corporation (ABC). Besides the difficulty of supplying accurate tuning devices to the ABC,[7] the Corporation's project to install tone generators in broadcasting studios and public concert halls, as suggested by the BSI, proved overly ambitious. At £500 apiece, tone generators were far too expensive for the ABC's budget; and the first of these devices was only installed in 1960.[8]

The prospect of implementing concert pitch through radio seemed even more unlikely in less industrialized countries. For example, in their response to the questionnaire sent by the BSI after the war, musical authorities in Mexico declared that although "it would be commendable" to use radio transmissions as a means to secure the adoption of the standard, "the Committee believes that it is very difficult to carry out in practice."[9] Reliance on radio as a tool for implementing *A* 440 meant the world's sonic integration was increasingly dependent on countries' wider stage of industrial and technological development.

Economic and technical struggles were not the only obstacles. As French musicians' opposition to ISO would demonstrate, and like decades of pitch negotiations had made clear, enforcement of the 1939 decision required, beyond the creation of an abstract norm and material prototypes of the standard, the cooperation of musical parties. Other standards could be established through the publication and circulation of the recommendations of the relevant professional networks; in the case of concert pitch, this was not sufficient. Standardizers needed to convince the musicians themselves to adopt the new measure.

Another, wider threat to the standardization process was the lack of public awareness of or interest in the standard's existence. A 1947 incident within the BBC itself illustrated this point. As the BBC engineer Francis William Alexander complained to one of his colleagues in the Corporation's Talks Department, a comment made by the acclaimed writer and physicist Edward Andrade on 31 January during a show entitled "Sound on the Air" was "really unfortunate": Andrade, apparently, had declared that "concert pitch has now the value for centre C of 256."

The lack of public awareness of ISA's 1939 resolution was a problem worldwide. In 1950 in Melbourne, for example, the newspaper *Argus* reported that *A* 440 was in general use in the city, but mistakenly attributed the introduction of this standard to the singer Melba, who had contributed to the adoption of the diapason normal in 1909.[10] In 1953, a German letter to ISO reported that, in response to the BSI inquiry, the new Physikalisch-Technische Bundesanstalt (PTB) had circulated 150 copies of a questionnaire "among a number of orchestra leaders in order to determine the degree of interest of practicing musicians in the maintenance of a certain musical pitch," and that only thirty had responded to their inquiry. "The exceedingly small percentual total [*sic*] of replies," the letter went on, "indicates that the general interest of those most concerned in [*sic*] the problem of musical pitch is very small indeed."[11] Already in 1941, an American study had shown that "the response to a broadcast frequency of 440 was rather disappointing." This publication contributed to the conviction of the chairman of the BSI pitch committee, Lloyd S. Lloyd, that one needed to make "more propaganda" about the standard.[12] His efforts to make concert pitch audible illustrate the difficulties arising from the social engineering of concert pitch.

The BSI was eager to use the BBC as a window for the standardization process. In addition to the broadcast of orchestral performances using the new pitch, they hoped to use broadcasting as a propaganda tool to forge musicians and instrument makers' adhesion to the new measure. Most scientific and technical standards were prized precisely for their invisibility. In contrast, a standard pitch required advertising strategies to trigger the cooperation of broad portions of society.

The shortage of money and manpower at the BBC in the aftermath of World War II had limited the plans for implementation of the 1939 resolution. Although at the end of the 1930s the Corporation had agreed to equip all its studios across England with a standard tuner, and other broadcasting companies in Europe had long before taken such measures, the BBC had to limit itself initially to simpler and less expensive measures, including the broadcast of *A* 440 signals on the radio and the supply of this tone to a single studio in London.[13] In addition to this technical program, the BSI asked the BBC to orchestrate a propaganda campaign about the reform. The question was: what did it take to make radio listeners hear concert pitch *A* 440?

The BSI pitch committee reconvened for the first time since the war on 2 July 1946. During this meeting, the members of the body agreed on several resolutions intended to publicize the creation of concert pitch. Since it seemed likely that "the general broadcasting" of the tone on the radio "might not occur for some time," the committee first resolved that "it would be of considerable value for the B.B.C. orchestras to have the standard note sounded before each performance."[14] For this purpose, the committee detailed that "a tuning fork closed in a resonance box and bowed, was the best method" and that "the note should be sounded for some minutes during the tuning up and should be audible to the public during the time when the tuning up was heard." As Pawley later explained to his colleagues at the BBC, the intended goal of this measure was "to bring the tuning of the orchestra to the public's attention," as well as having "the general public [appreciate]" the fact that the BBC orchestra conformed to the standard. During the following meeting of the committee, the music director at the BBC, Mr. Biggs, believed that it "might be helpful . . . to permit the public to hear the tuning up of an orchestra to the standard note," as well as "have considerable educational value and encourage the use of the standard note in other concert halls and by orchestras other than those sponsored by the B.B.C."[15] Despite Biggs's support, technicians at the BBC met the BSI pitch committee's suggestion with only moderate enthusiasm. In March 1947, after discussing the matter with various services at the Corporation, Pawley reported that tuning up could not become "a general practice," but "might be done occasionally."[16]

The BSI pitch committee was eager to implement its prewar decision to regularly broadcast *A* 440 frequency signals outside of musical performances, and then further tune the ears and minds of British audiences to the reform. Pawley made as much clear during an exchange with the BBC superintendent engineer that took place shortly after the first meeting of the BSI's pitch committee. To his colleague's observation that the broadcast of *A* 440 was likely to take place "before 6.30 a.m.," when neither musicians nor makers were likely to be at work,[17] Pawley's response was to emphasize the measure's dual function: not only that musicians and instrument makers could "check their own tuning fork or other sub-standards of pitch," but also, mostly, "to arouse interest in the desirability of adopting a standard of pitch."[18] In other words the BBC, by broadcasting this tone, was addressing a social as much as a technical need.[19]

It was decided, finally, that the standard tone be used "before the beginning of the Third Programme each evening," at 6 o'clock. The emission of this tone introduced a significant change in the BBC's radiophonic soundscape: it indeed replaced the broadcast of a 1,000 hertz line-up tone that listeners could use to align their radios. Yet the committee believed that the measure would not "speak" for itself: to become a valuable asset in the BSI's propaganda campaign, the broadcast required some verbal explanation. This is why, during their 2 July 1946 meeting, the committee members resolved that "in order to familiarise the public with the idea of the standard note . . . , a short talk should be broadcast calling attention to the new development, and explaining the reasons for broadcasting the note."[20] As plans to implement the diffusion of this tone were developing, the BSI and the BBC resolved to air a program intended to educate the public about their reform.

On 29 May 1947, the BBC transmitted a segment titled "Can We Have an 'A'?" The program brought together several of the authorities who had been involved in prewar negotiations of a standard pitch to alert the public of the change in their radiophonic soundscape and to announce the implementation of the international agreement made in London eight years before.[21] It started with these words:

> ANNOUNCER: This is the BBC Third Programme.
>
> Tomorrow we shall make a slight change in the presentation of this service by altering the tuning note we transmit before the programme opens at 6 o'clock.
>
> The note we shall be using is "A" at the International Standard of Concert Pitch, which was agreed for use in most countries in 1939, but wasn't put into practice immediately because of the war. We intend to transmit this note every evening, because we believe it will be a useful datum for people responsible for musical performance, for instrument makers, for tuners, and everyone else who's interested.[22]

Initially envisioned as an interview with Lloyd, chair of the BBC's pitch commission, about the work of the BSI pitch committee, the show in its final version turned out to be a rather contradictory discussion among representatives of the diverse categories of actors involved in the making of concert pitch: Pawley, the wind instrument maker and veteran of the 1939 conference; the wind instrument maker Arthur Blaikley; the BBC orchestra conductor Stanford Robinson; the violinist Margaret Duff-Challen; and the oboist John Wolfe. During the show, the latter two played the role of "perfect pitch representatives."[23] Instead of providing a single, straightforward explanation about the standard, the program gave listeners a vivid sense of the complexity characteristic of concert pitch by exhibiting the plurality of conceptions that continued to surround it.

With its emphasis on the variety of social groups and opinions, "Can We Have an 'A'?" may have in some ways seemed to reenact the negotiations that had taken place in 1939. Yet the conversation also involved participants who had not taken part in these debates, including the three musicians from the BBC.

During the show, the respective time allotted to the various parties was inversely proportional to their authority in the debates. After Pawley briefly detailed the level of accuracy of the new tone, the conductor Stanford Robinson took the floor to explain how "once you've given us an international pitch that's where our troubles begin."[24] To begin with, "much of our work's done in places where the temperature's different from that selected by the makers when they design the instruments." For example, Robinson recalled at "the Albert Hall this past winter," "the thermometer registered 45° one morning."

In such situations, the conductor asserted, "the players . . . have to adapt their minds to quite a varying set of conditions, and it isn't fixing an international standard that makes them play in tune."

In addition to pointing to the difference between a sound engineer's production of a standard signal at the BBC and musicians' practices, Robinson further outlined the divisions between different categories of instrumentalists, thus depicting the orchestra as a battlefield of opposing approaches to pitch, and the site of intense negotiations between these various parties. As he declared: "One of the biggest problems in the orchestra is to stop the eternal dogfight between the strings and the winds." The conductor further explained how "The winds have a firm conviction that the strings play sharp and they're often absolutely beastly about it—they really are. The strings (of course) say they only play sharp to try and keep up with the winds which are rocketing through the performance and won't stay down."

This was not the only reason why one needed a musical standard. Musical pitch's great instability indeed not only created conflicts within orchestras, but had a detrimental impact on musicians' ears. Robinson explained: "the value of the standard . . . is that in the course of a few years or so, a young generation of musicians will come along who by sheer habit will have developed an instinct for accurate pitch." Duff-Challen agreed, adding that "everyone ought to have the chance of *growing up* with one definite pitch. I think many people would have perfect pitch if they hadn't been confused during their musical training by having to play and listen to so many different pitches."

In giving these actors a voice that they had not had in pitch negotiations, one could argue that "Can We Have an 'A'?" was a mock debate that inaccurately portrayed the power relations that had underpinned the standard's creation. But by going beyond technical questions and turning to musicians' views on the standardization process, the show actually provided listeners with a realistic picture of the reform. After all, beyond the BSI and BBC's joint technical engineering of the standard, musicians now had a considerable part to play in the manufacture of concert pitch. In this way, the show epitomized the balance of power that presided over the making of the standard.

It is difficult to determine the impact of the BBC's propaganda on British musical communities and on wider society,[25] but there is some evidence that it was slight. On 29 April 1949, for example, Pawley complained that J. Raymond Tobin's 1948 book entitled *A Seat at the Proms* mistakenly stated that "what we call 'concert pitch' or New Philharmonic pitch is fixed by the A to which the orchestra tunes being the result of 439 vibrations per second (v.p.s.) at 68° Fahr." Concluded Pawley: "It appears that the publicity given to [the standard] has not permeated as far as we had hoped."[26] The question of concert pitch's audibility was still a subject of conversations between the BSI and the BBC in the winter of 1949–1950.[27]

TO CERTIFY OR NOT TO CERTIFY? MUSIC AND THE LIMITS OF TECHNO-SCIENTIFIC UNIFORMITY

Besides agreeing to the frequency *A* 440 at the 1939 London conference, participating countries had agreed on a list of questions designed to guide their efforts in implementing the standard on the national and international level. Though most countries had deferred their inquiries due to the war, Italy, Germany, and Norway had started to review this program after the outbreak of hostilities. Thus, by the time the BSI started to organize conversations about musical pitch in 1945, the results of several initial studies were already available. On reopening international pitch negotiations, the BSI re-dispatched the 1939 questionnaire to ISO members: in response, further studies followed in other countries, including France, England, Australia, Austria, Hungary, and Mexico. This research introduced a new perspective to the debates.

Several experiments combined traditional physical approaches with new insights from psychology, musical acoustics, and psychoacoustics. Some of these studies on musical pitch brought musicians into the laboratory and, conversely, transformed musical venues into sites of scientific inquiry. For instance, psychological factors played a part in the production of musical pitch. Several studies demonstrated not only that musicians played sharper in emotionally tense passages, but that pitch was itself a subjective phenomenon formed by the interplay between frequency and other parameters, including loudness and quality of tone. Given musicians' agency in the making of the standard, standard-setting bodies refused to take full responsibility for the standard, issued ambiguous categorizations, and thus left the definition of pitch open to the variety of its musical appropriations.

One of the first studies to be conducted in the aftermath of the 1939 London conference had taken place in 1940, at the Italian National Institution of Electro-Acoustics (Istituto nazionale di elettroacustica), where ethnomusicologist Ottavio Tiby (1891–1955) and physicist Alfonso Barone conducted a study of pitch variations during the performances of what the authors described as "the great Italian theatres and among the important Italian concert organizations," including the Royal Opera Theatre of Rome, the Scala Theatre in Milan, Verdi Theatre in Trieste, the Carlo Felice Theatre in Genoa, and the Ente Italiano per le Audizioni Radiofoniche in Rome and Torino.[28] Likewise, between September 1939 and January 1940, at Bell Laboratories, O. J. Murphy had led a series of measurements during 750 concerts broadcast on American radio stations using the same apparatus and method as van der Pol in 1938.[29] Similar experiments kept developing after the war. At the new PTB established in Braunschweig, for example, Lottermoser and von Braunmühl continued to carry out measures on the frequency of musical pitch in broadcast performances, in collaboration with the radio station Nordwestdeutscher Rundfunk (NWDR), only with a "different apparatus" than before the war. Just as in the PTR's 1930s experiments, these new observations revealed that "the pitch is still without exception fixed too high by several c.p.s."[30]

These post-1939 studies brought new perspectives to the pitch debates. Reflecting the taxonomies in use in American radio programming, for example, Murphy's 1939–1940 article provided a comparison between different musical ensembles, including "Symphony orchestras," "Light orchestras," "String groups," "Organs," "Pianos," "Misc[ellaneous] solo inst[ruments]," "Brass bands," and "Dance bands." As Lloyd emphasized after reading this publication, the most striking outcome of this study was string players' lack of observance of the standard. In comparison, brass bands, although constantly accused of raising the pitch, were far less problematic.[31] In Europe, scholars sought to expand the scope of investigation by including performers' views and practices. In Germany, for example, the PTR's measurements were "supplemented by a questionnaire addressed to the orchestra leaders and bandmasters." The outcomes of this study presented not only statistical data, but a sampling of individual approaches to the problem.[32] For instance, the authors reported on an orchestra whose "principle was to cultivate only works without wind instruments since their musical pitch tends to rise," as well as another whose "answer to the question of who adjusts [to] whom" was "the tuning fork of the piano tuner . . . to the solo-oboe."

Tiby and Barone's inquiry, completed shortly after the outbreak of war, exhibited the same shift from the measurements of frequency to an examination of musicians' practices. As the authors explained, they were convinced that it was not enough to describe the problem of variation to fix it: rather, one needed to explain it. To do so, Tiby and Barone explored a wide array of questions, ranging from aesthetics and music theory, to history and psychology. Conducted ten years before the German study, under the fascist government of Mussolini (it was funded through the Ministry of People's Culture), the study had a nationalistic and aesthetically conservative character radically at odds with the PTR's approach, which reflected West Germany's eagerness to develop an inclusive and democratic approach to the standardization process. Despite its problematic celebration of the nineteenth-century orchestral canon and its "glorious" contemporary Italian exemplars, Lloyd read it with the greatest enthusiasm and circulated it widely among standardizers. This Italian study thus exerted considerable influence over postwar pitch conversations.

Tiby and Barone wanted to examine the behaviors of instruments by observing specific notes played by specific groups within the orchestra.[33] Instead of measuring broadcast performances from their laboratories as Lottermoser and van der Pol had done, they placed recording devices within the orchestra that registered specific localized *A*s that the orchestra had previously selected. They also conducted complementary observations outside of any musical performances, by recording various players performing short phrases or single notes. Drawing on their observations, Tiby and Barone described the myriad sources of frequency variations in musical performance, ranging from instrument design to temperature, to performers' voluntary or involuntary actions. To begin with, the scholars emphasized the great variety of tuning systems and, in particular, the immense differences between equally tempered instruments such as the piano and the organ (at least, the most recent ones), and wind instruments embodying the natural scale, such as the cornet.[34] A similar observation came in the response to the 1950 BSI inquiry of Mexican composer Julián Carrillo, who emphasized that "perfectly tuned wind instruments do not exist," and that "the whole of the instruments of symphony orchestras are in disagreement with the temperate [*sic*] system." To exemplify these tensions, Carrillo reported on a controversy that occurred between a pianist and a violinist as they were about to perform Bruch's concerto in G minor:

> The pianist was protesting because the violinist's Sol appeared to sound too flat, to which the violinist replied that he had just tuned his instrument. Both were right, since the piano Sol did not coincide with the violin Sol; owing to the fact that the violinist had tuned to the fifth RE-LA in physical fifths, which are higher than the temperate fifths, for which reason the RE was already flat, and on that RE flat, he tuned the Sol, for which reason this was twice flat.[35]

Was the lack of uniformity among instruments' tuning systems actually a drawback? According to Tiby and Barone, this diversity constituted the essential beauty of music. As they wrote, "Just as an instrument whose sounds deprived of harmonics would have a saw-like and cold timbre, and would thus not be artistically useful, so an orchestra constrained to absolute precision of sounds would be quite other than musically satisfying."[36]

Other causes of pitch variations identified by Tiby and Barone, however, were not so desirable, in particular those they presented as involuntary causes: temperature changes and performers' lack of control over intonation. Nevertheless, intentional variations like tuning systems could be a source of aesthetic satisfaction. As an example, they cited the vibrato of strings, "used to give the colour and accent of passion to the sound."[37] They further claimed that there were many other "artistic," or "emotional," causes for pitch variations:

> it would seem useless to dwell on the constant occurrence of a phenomenon which, although necessarily noted, has always been obvious in the musical field. A "sforzando" of the sound, a "fortissimo" (especially if taken in the upper register), a more intense expression, a specially [*sic*] vibrant execution, are other causes that may produce a slight and unconscious, but nevertheless effectual, increase in the frequency of sounds.[38]

Passages throughout Tiby and Barone's study made plain the impossibility of reconciling music's history, theory, aesthetics, and material culture with science and industry's demands for uniformity. Fixity in relation to musical performance was a "utopian" notion. Instead, Tiby and Barone celebrated variety:

> if, per hypothesis, it were possible for us to maintain the orchestral performance at a rigorously constant level of pitch, that is if it were possible to fix the sounds that are of course perpetually mobile, suppressing all the causes voluntary and involuntary of instability, we should be rendering music the worst possible disservice; since not only would we make suddenly manifest all the differences in intonation of the different instruments . . . and would thus have in the orchestra a continual source of noticeable discords, but we would also have forbidden musical expression, colour and liberty of execution . . .[39]

Drawing on these aesthetic claims, Tiby and Barone called for a flexible definition of the pitch standard. Given the desirable variations indicated by their experiment, the two authors demanded that "during the performance, a tolerance of plus or minus 2 c.p.s. . . . be allowed," in other words, that standard pitch be defined as a range of four frequencies between 438 and 442.

As much as Lloyd praised Tiby and Barone's report, he was skeptical about this last point. Engineers, he argued, would never agree to such an imprecise definition of the measure. Yet he took full note of the implications of the Italians' findings when writing that "the utmost which could be achieved by standardization was to ensure that the orchestra was initially tuned correctly, by giving them an accurate standard note to which to tune." "The maintaining of pitch during a performance," in contrast, "was in the hands of the conductor and the players themselves."[40] This observation was of considerable significance for the negotiations. Crucially, it put concert pitch at odds with existing definitions of industrial standards. Standards agencies manifested the conformity of manufactured objects to their norms not just with the publication of written specifications, but through the grant of certification marks. Accepting the agency of musicians in the production of concert pitch, however, meant that this standard would not match the criteria that such procedures required, and that the implementation of the measure would not follow the same path as that of other standards. Industrial actors may have shaped the final stage of pitch negotiations, but concert pitch had in turn transformed these actors' approach to the process. The adjustments in the debates made by the ISO and other standard-setting bodies revealed just how restricted the territory of application of their regulations would be.[41]

Murphy, and Tiby and Barone, in revealing the impact of musicians on the new standard, also demonstrated the limits of techno-scientific authority over music. This became clear in standards agencies' conversations surrounding the questions of certification. The BSI tackled this point in their late 1940s conversations when discussing edits of the standard issued in 1939. On 31 January 1949, they reviewed the conditions governing the use of BSI certification marks on products, which "certify that [they] complied with the requirements of a particular British Standard," noting:

> one essential requirement of any license to use this Mark would be that the manufacturers should have in operation a scheme of routine inspection and testing to ensure that all marked products would comply with the requirements of the British Standard, and that independent check tests on marked products would be made by the B.S.I., from time to time.

Blaikley observed that "it would not be practicable to use the B.S.I. Mark on instruments to certify compliance with [the standard]," arguing that "a note obtained from an orchestral wind instrument would depend not only on the design and manufacture of the instrument, but upon the player" and, as a result, "it could not be assumed that there would be agreement" between tests performed with different people. Familiar as they were with the question of musicians' agency through their discussions of Tiby and Barone's paper, the other members of the committee agreed with Blaikley and renounced the idea of certifying the conformity of instruments with concert pitch.[42]

AFNOR similarly rejected the idea of issuing certification marks for musical instruments. After AFNOR's technical director Charles Duval "emphasized the need for a certification mark" during the first meeting of the organization's pitch committee, claiming that without it the standard "would lose some of its interest," the wind instrument maker Acoulon pointed out that such certificates could only apply to keyed instruments such as the flute, the oboe, or the clarinet. At a following meeting, however, Acoulon declared that he doubted the utility of such a practice altogether, since "makers sign their instruments and, thus, take responsibility for them," adding that "an instrument like a clarinet can vary by more than a quarter of a tone when the performer changes."[43]

THE "PITCH QUARREL": FRENCH MUSICIANS VERSUS ISO

Eight years after the head of the BSI first heard of potential French disgruntlement with the 1939 London pitch agreement, delegates from seventeen countries met in London and renewed the choice *A* 440 as the international tuning frequency, despite the stubborn opposition of several of the major French state cultural institutions. The French crusade against *A* 440 was spearheaded by the musician Robert Dussaut (1896–1969). On learning about the 1939 agreement reached at London, which had been signed by the French standards agency, AFNOR, Dussaut launched a campaign to denounce what he characterized as a problematic compromise with the American music industry and Nazi Germany. Instead of *A* 440, the composer promoted the lower frequency of *A* 432 that had been endorsed by several parties at the end of the nineteenth century. Dussaut was a "survivor of a disappearing world," a former student and a solfège teacher at the Paris Conservatory at a time when France's most promising composers had deserted the institution, and a composer of works pursuing the nineteenth-century tradition of *opéra comique*, oratorio, and symphony.[44] The composer's interest in the musical past expressed itself through his research into the history of tuning systems; his rejection of *A* 440 epitomized the anxieties of France's most conservative musical circles faced with the decline of their own nation and cultural values. In the wake of Germany's occupation of France and the development of industrial modes of global governance, Dussaut's campaign attempted to restore the authority of the classical music artistic community of France against what he perceived as the problematic imposition of foreign and commercial diktats.

Dussaut made contradictory claims about the start date of his campaign against *A* 440,[45] but the composer's initiative certainly took off in February 1950, after AFNOR received a letter from the BSI inquiring about the organization's efforts to implement the standard adopted in 1939.[46] On 25 February 1950, Dussaut held a meeting to discuss "the fixation of a new French pitch" at the Institut International du Son, an organization he had recently co-founded to foster the "documentation, centralization, and diffusion of acoustical research."[47] During the following months, the composer led what he called a "Referendum," gathering letters from musicians expressing their support of his proposed reform.[48]

To try to change the course of pitch negotiations, Dussaut mobilized diverse networks. He first presented the results of his efforts in front of AFNOR. In March and April 1950, the composer attended two meetings of a "select acoustic committee" that the organization had gathered to discuss the BSI's request. Attendees to the first meeting included Charles Duval, AFNOR's technical director; José Bernhart, a sound engineer at the Radiodiffusion and a veteran of the 1939 conference in London; and Jacques Couvreur, the general delegate of the national federation of musical industries and commerce; as well as the wind instrument maker Alfred Acoulon, and the engineer and acoustician Robert Cabarat.

This meeting made it clear just how isolated Dussaut was in this arena, where representatives of the music industry were overwhelmingly in the majority. The composer opened the discussion by expressing his main grievance about the 1939 decision: "that musicians had not been consulted in 1939 because they would have opposed the adoption of the frequency of 440."[49] Duval rejected this objection, pointing out that Henri Rabaud, the Paris Conservatory's director at the time, who had just died in 1949, had signed the London agreement. For Dussaut, however, this signature was simply a testimony to AFNOR's "bad faith."[50] According to the composer, Rabaud's signature had been obtained in Paris "by surprise," and after the fact. "Rabaud was not invited to the Congress's deliberations," Dussaut claimed, and had not had time to consult with any representative of his profession. Thus, his signature "was not made in the name of the collective of all French musicians."[51] By contrast, Dussaut alleged that, thanks to his referendum, he was in possession of "hundreds of signatures, and thousands of adhesions all over France" in favor of *A* 432 and, thus, in the position to testify that "musicians in general disapproved of the *A* 440."[52]

Responding to Dussaut, Duval not only rejected the idea that musicians had not had a voice in 1939, but also predicted that no nation would agree to revisit the decision to adopt *A* 440.[53] Quoting a study published in the Bell Telephone System's technical bulletin, Duval reported that "the average frequency establishes itself at 441.3 periods per second for all orchestral and solo concerts."[54] Couvreur agreed with Duval, adding that "since 1939 all the tools were adapted to the frequency 440 and that a new change would involve considerable expenditure."[55] After the meeting, Dussaut shared his anger and distrust with Claude Delvincourt, the director of the Paris Conservatory who had succeeded Rabaud, declaring that he was the "pet peeve" of wind instrument makers and broadcasters, and calling Acoulon and Bernhart his "adversaries."[56]

A few weeks later, AFNOR held a second meeting during which Dussaut was beaten at his own game, with several members of the body mobilizing musicians' authority to counter the composer's claim that his profession disagreed with the 1939 decision. An engineer of the firm Couesnon displayed letters from professors of the Conservatory who asked that "their names be crossed off the referendum." Confronted with this evidence, Dussaut replied that "they were all testers for the firm Couesnon," and that it must be the maker "who order[ed] them to retract themselves and provoked these letters."[57] Similarly, Bernhart introduced two famous performers, including the authoritative conductor of the Radiodiffusion orchestra Désiré-Émile Inghelbrecht, who defended the BSI's and AFNOR's suggested standard. Although Inghelbrecht agreed that "one needed to create an atmosphere of decrease," he nevertheless observed that "in Bach's time, for example, instruments were not the same as today and instrumentalists were not as good as today. Pitch's escalation thus responded to a certain need."[58] After attending AFNOR's two meetings, Dussaut declared himself convinced that the organization, the Radiodiffusion, and wind instrument makers were leading a conspiracy against musicians,[59] a situation all the more problematic since, according to Dussaut, they "don't have a musical ear."[60] As P.-J. Richard, the president of the Institut International du Son, put it in a letter of support to Dussaut: "to adopt the degrees of the thermometric scale, did we ask thermometer makers for their opinion?"[61]

Dussaut's attempt to employ alternative networks to counterbalance the private industrial actors in pitch debates failed. After the meetings at AFNOR, the composer established contact with the institutions that in the nineteenth century had orchestrated France's initial efforts to unify pitch: the Academies of Fine Arts and Sciences, as well as the Ministry of Education. His initiative resulted in the revival of the state-driven approach to pitch standardization that France had developed a century earlier. This strategy did not prove successful, however.

On 9 April 1950, Dussaut sent a *Discussion on musical pitch* (*Discussion sur le diapason*) to the Academy of Fine Arts,[62] a text full of political suspicion about the adoption of *A* 440, presenting the 1939 decision as a commercial diktat. Dussaut claimed that "Dr. Grutmacher [*sic*] and the acoustic committee of the Berlin Radio" had been the ones who, "in September 1938 (at the time of Munich!)," asked the BSI to organize a Congress in London so that their own standard would be adopted internationally.[63] The references to Nazi politics were an attempt to strike a patriotic chord and allow Dussaut to paint his reform as a way of restoring the authority and honor of France after its brutal occupation by Germany.

Dussaut laid the blame for the 1939 decision on a second, radically different category of actors: American jazzmen. As he wrote, "our musicians and our singers [are] dependent on jazz players from across the Atlantic." Dussaut's denunciation of the empowerment of the United States and popular music showed that the "pitch quarrel" went far beyond physiological and cultural concerns over the conservation of voices and historical musical works. It also carried cultural, political, and racial implications. On 1 June 1950, for instance, the critic René Dumesnil published an article in *Le Mercure de France*, in which he praised Dussaut's initiative, accusing "negro trumpets and saxophonists" of "imposing on old Europe a pitch thanks to which our singers scream and the treasures of our symphonic and dramatic music become inaudible."[64] Similarly, in October 1950, the journalist René Sudre lamented in the *Revue des Deux Mondes* that *A* 440 "has become the universal norm . . . thanks to the contagion of the [United States'] Negro music."[65] For Dussaut, Dumesnil, and Sudre, musical standardization crystallized deep political, aesthetic, and racial anxieties. As Dussaut wrote to the Ministry of Education, if France were to adopt *A* 440, then the nation might as well "give up the franc, the meter, the French language."[66]

Along with making musical pitch into the instrument of a flurry of nationalism and racism, Dussaut's *Discussion* justified the composer's suggestion to adopt *A* 432, citing the history of both music and science, arguing that "between the current *A* (at 440) and that of Bach's time (at 405), there is almost one tone," and that "the frequency 432 imposes itself, because this number is Pythagorean," and in using it as a point of reference, "one obtains . . . a chromatic scale whose frequencies can all be expressed as whole numbers." Dussaut claimed to have broad support among musicians, declaring that he had collected the signatures of 23,000 of them, including: Delvincourt "and almost all the professors of the Conservatory in Paris," Dussaut's former teacher and friend Henri Büsser "and almost the totality of conductors and choir and singing directors, singers, and musicians of the Opera," as well as Paul Bastide "and the whole Comic Opera: the 70 musicians of the orchestra, and all the conductors, singing directors and singers."[67] Dussaut presented himself as the spokesperson of French musicians, demanding that the minister of education declare the 1939 agreement "null and void" and "organize an official French congress to adopt a new French pitch at 432." In a direct reference to the commission that had established the diapason normal almost a century earlier, Dussaut continued, "musicians should be in majority, as in 1859."[68]

The members of the Academy responded favorably to Dussaut. On 21 June 1950, they unanimously agreed to write to the minister of education and demand that "a commission composed of musicians and physicists be charged to fix in a definitive way the level of musical pitch." Following this favorable outcome, and seeking more support from French traditional authorities, Dussaut then submitted his proposed reform to the Academy of Sciences.[69] Though drawing largely from the *Discussion*, this *Note* put more emphasis on the scientific arguments that were most likely to arouse his audience's interest. Through the *Note*'s title, "Proposition of a New Fixed Tone," he inscribed his proposed reform in the tradition of the proposition made by Sauveur to the very same Academy in 1700, when he suggested the adoption of a "fixed sound."[70] Although Dussaut praised Sauveur's selected measure ($C = 256$), he explained that it was not a convenient standard for musicians, "because violinists can only tune on their open strings." "The fixed tone," Dussaut asserted, "must thus be only *A*, *G* or *D*, these three notes corresponding to the strings of the violin, the viola, etc." While "for a long time, musicians have adopted A_3 as a standard," the composer continued, "this choice has been very unfortunate, for *C* then becomes highly variable: the minor third from A_3 to C_4 differs a lot depending on whether it is Pythagoras' system, [equal] temperament, or Zarlino's system." To overcome this difficulty, Dussaut suggested that musicians adopt *G* as a point of reference, from which one would obtain "*both in Zarlino's and in Pythagoras' system* the same natural fifth . . . from *C* 256 to *G* 384, as well as the same natural fourth . . . from *G* 384 to *C* 256."

The Academy of Sciences welcomed these arguments with enthusiasm and, like the Academy of Fine Arts, agreed to support his proposal to the minister of education.[71] Jacques Jaujard, the director of arts and letters at the Ministry of Education, responded positively in turn, and on 6 August 1951 appointed a commission in charge of studying the question of pitch unification. This body turned Dussaut's wish into a reality, in that it was largely dominated by musicians, like the commission that had created the diapason normal in 1859.[72] The overrepresentation of musical interests in the negotiations had been controversial already in the nineteenth century. In 1950 their argument was equivalent to a militant gesture aimed at combating what Dussaut and his allies conceived of as the unbearable takeover by industrial actors.

The appointment of this commission seemed to mark an important victory for Dussaut. The inadequacy of the musicians' authority, however, quickly became apparent. Dussaut's revival of a nineteenth-century approach to pitch standardization revealed just how little power national governments and traditional musical and scientific authorities and arguments had in mid-twentieth-century pitch negotiations. While the ISO continued to cultivate the technocratic internationalism that had begun before the war, the French government was trying to resuscitate a diplomacy of sound typical of the 1850s. Yet Dussaut and the state-appointed commission had no impact whatsoever on the negotiations, confirming in fact their fear of losing all influence. Dussaut himself recognized as much in 1954, noting that "since its creation, [the commission] has done absolutely nothing."[73] The anachronistic character of Dussaut's initiative was already clear in October 1953, by the time the ISO was preparing to hold the conference on musical pitch in London. On learning that the BSI was making progress with the organization of this event, Dussaut convinced his allies to stage an intervention. It was an ambivalent move: by addressing the ISO, Dussaut was implicitly acknowledging the organization's authority over pitch matters. The reverse was not true. The composer learned that, as far as the international federation was concerned, AFNOR was France's only valid spokesperson on pitch matters—pitch diplomacy was firmly in the hands of standards agencies.

On 8 October 1953, on learning through Dussaut about the ISO's congress,[74] Delvincourt suggested that Dussaut write a letter stating that "the conclusions of the . . . London Congress won't have any value in musicians' eyes in general, and in those of French musicians in particular," and emphasizing the purpose of their intervention: to protect musical works from the "mutilation" that resulted from pitch's escalation.[75] This letter, Delvincourt concluded, should display the signatures of several French authorities. Following Delvincourt's advice, Dussaut drafted a "letter of protest" on behalf of the government's secretariat of fine arts, and managed to have it signed by Delvincourt, Jaujard, and other members of the state pitch commission,[76] as well as Maurice Lehman, the general administrator of national theaters, and Ernest Esclangon, the president of the Academy of Sciences.[77] In this letter, which Dussaut sent to both the BSI and AFNOR, the composer warned the two organizations that "*French musicians and physicists will not accept the coming decision of the London Congress.*"[78] As Delvincourt had predicted, this initiative did not have any impact on the conference's deliberations. The 1939 decision was validated, and the decision first published in 1955 as ISO Recommendation R16 (and then, finally in 1975, as ISO 16). In Dussaut's words, "the Fine Arts Secretariat's intervention [was] considered to be nonexistent."[79]

Dussaut's intrusions did not go unnoticed. After the congress, Jean Birlé, AFNOR's general director, complained to Delvincourt that Dussaut's letter had called the attention of the international community to the lack of coordination between France's musical circles, a development that "had been very prejudicial to the authority of our delegation," according to Birlé.[80] This initiative was all the more unfortunate, Birlé asserted, in that "most of the people who signed the collective are members of AFNOR's acoustic commission."[81] Birlé concluded that to avoid such confusion in the future, the ISO's secretary had recommended that French musicians attend AFNOR's meetings.[82]

In addition to blaming the manner of Dussaut's intervention, Birlé contested his reform itself. Calling attention to "the results of several referenda" presented in London, Birlé added that he was convinced that "many musicians consider that the old frequency 435 Hz does not respond to current artistic demands anymore."[83] Delvincourt asked Dussaut to draft a response to Birlé.[84] In this letter, Dussaut reasserted the main points of his argumentation, concluding that "we must go back to the principle admitted in 1859, that is let musicians and physicists make their own choice for the best frequency of A_3."[85]

REGULATING AUDIENCE AND PERFORMERS: THE SUBJECTIVITY OF PITCH

At the same time that Dussaut was mobilizing nineteenth-century scientific networks and arguments to critique the new standard, mid-twentieth-century conversations on pitch became increasingly entangled with the conceptual and technological tools of psychoacoustics, developed within applied research on speech and hearing. This added a new layer to the unification program that standardizers pursued. In their search for precision, actors involved in the debates increasingly distinguished between the physical, objective phenomenon of frequency and its psychophysiological, subjective counterpart: pitch. Since the interwar period, several experiments had highlighted the gap between sound waves and their perception by humans. Drawing on this new knowledge, acousticians began to call for a clear demarcation between frequency and pitch in conversations surrounding musical standardization. For example, the American acoustician Robert W. Young, from the US Navy Electronics Laboratory in San Diego, the United States' spokesperson in postwar pitch negotiations,[86] noted how "in these discussions of concert pitch it is a little awkward that one is really specifying a standard of frequency and not pitch."[87] Displaying the same concerns as Young, more and more actors started to mobilize the psychological definition according to which pitch was "that attribute of auditory sensation in terms of which sounds may be ordered on a scale extending from low to high, such as a musical scale."[88] In 1946, the American acoustician Harvey Fletcher further detailed how, far from being located in the waveform, pitch was rather to be found in the ear, "probably related to the position of the stimulated nerve endings on the basilar membrane of the ear."[89]

To be sure, acousticians involved in the debates had long been aware of the psychophysiological implications of their observations, and, as shown in the previous chapters, during the nineteenth and early twentieth centuries, research on musical pitch had developed in close proximity to studies on tone perception and discrimination.[90] Yet until the Second World War, conversations about musical standardization mostly equated pitch with frequency. Things changed with the empowerment of industrial actors in the late 1930s, particularly as standardizers started to think about the implementation of concert pitch after the 1939 London conference. Integrating musicians into the conversation about pitch standardization meant not only considering questions of musical performance but also addressing matters of auditory perception. As Lloyd put it, it became increasingly important to understand not only "what . . . the violins and the wind instruments do" when tuning to A 440, but also what "they felt." As Harvey Fletcher explained in a 1946 article emphasizing the need to distinguish between pitch and frequency:

> Because of its subjective character the measurement of pitch can be made only by judgement tests. Since individuals vary in such tests, it is necessary

> to use several listeners and average the results in order to get something that will be characteristic of a "typical" listener.[91]

Fletcher's statement captures a dialectic central to the field of psychoacoustics at the time, between the variety of auditory perception on the one hand, and the engineering of "normal subjects" on the other. In their attempt to maximize the efficiency of audio technologies (and, thus, the profits of telecommunication companies), American psycho-acousticians in the interwar period sought to determine the average range of hearing through the massive testing of listeners.[92] With the rise of psychology in pitch debates over musical pitch, the standard started to crystallize the same dialectic and serve as a disciplining tool not only for the unification of musical frequencies, but also for the formation of "typical listeners." But musicians were not mere passive listeners. In entering the standards' circuit of implementation, they played a disruptive part by revealing the highly individual character of auditory experiences.

Germans were among the first to implement the resolutions of the 1939 London conference. As mentioned in the previous chapter, by 1940 the Deutschlandsender was broadcasting *A* 440 signals every day at 10.45 a.m. and, in addition, engineers at the Reichs-Rundfunk had developed tone generators. It was hoped that these devices, initially installed in recording and broadcasting studios, would soon replace tuning forks and pitch pipes across the Reich's music venues, from dance halls and movie theaters, to operas and music schools. According to the artisans of these generators, Hans-Joachim von Braunmühl and O. Schubert, knowledge of auditory perception played a decisive part in their production, including the fixing of their standards of accuracy (of 3/1000 hertz), tone quality (a percentage of 25 percent of overtone content), and intensity.[93] In short, Schubert and von Braunmühl's tone generators embodied these actors' knowledge of human hearing, expressed in the notions of tone discrimination, timbre, and loudness.

After the war, the BSI pitch committee followed in the footsteps of this pioneering experiment. Just as in Germany, auditory research played a key role in calibrating the British standard. With an accuracy of "1 part in 10 million,"[94] as a BBC engineer humorously put it, the 440 hertz signal broadcast daily at the beginning of the Third Programme was "at any rate better than Sir Adrian's [Boult] ear."[95] More earnestly, when discussing the desirable level of accuracy for the calibration of tuning forks, on 31 January 1949, Lloyd cited Biddulph's screening of speech's auditory spectrum to determine "the smallest differences of frequency detectable in the average by the human ear," and

Carl Seashore's eugenic measures of musical talent "which had suggested that in the average . . . especially sensitive players would be unable to detect a difference of less than 1.25 c/s in 1,000 in the tones of their instruments sounding tuning-A."[96]

Auditory knowledge not only shaped the frequency accuracy of tone generators, but led acousticians to include related acoustic parameters in their design, especially loudness. Braunmühl and Schubert's article had framed this notion in terms of audibility, but postwar conversations about pitch standardization increasingly referred to it in relation to pitch. As Pawley explained to Alexander, "It is well-known that the pitch of low tones, including those in the region of 440 c/s, falls with increased loudness." This raised crucial questions regarding musicians' practices. "If the players tuned by beats," Pawley went on, "I do not see how this could happen; but presumably they tune by comparing the subjective effect of the tuning note with the subjective effect produced by their instruments." As a result, Pawley argued that "it is necessary to specify the loudness of the tuning note to make sure that the orchestra will tune correctly."[97] Such a notion carried far-reaching implications.

As these examples reveal, the inscription of musical pitch in psychoacoustics' conceptual and technological networks opened the door to increased scrutiny of musicians' perception and actions. Yet the relation between the measure and auditory research was not unidirectional. While knowledge of hearing shaped theoretical and material aspects of the standard, in turn, the implementation of concert pitch challenged established notions in the field of psychoacoustics. Calibrating tone generators to the ears of highly trained musical practitioners would eventually unveil the difficulty of producing universal laws about auditory perception.

While levels of accuracy and loudness were set a priori, the determination of the desirable amount of harmonics for the sound that tone generators supplying concert pitch emitted triggered experiments that ultimately revealed the reversible character of the relation between pitch standardization and research on auditory knowledge. In their 1941 article, von Braunmühl and Schubert had claimed that they had designed their tone generators according to musicians' preferences for harmonically rich tones.[98] Similarly, in Australia, in 1953, the national broadcasting corporation led a study whose conclusions were that "a pure tone is very unpopular with musicians" and that "[a] tone of timbre approximating that of the oboe is most preferred." This choice, the authors of the study believed, was "almost entirely due to the fact that in this country the oboe is still the commonest form of providing a note for orchestras to tune to."[99] By the time the first electric device tone generator was installed in an ABC studio, in 1960, the postmaster general's department had "worked to produce a waveform closer to what musicians were accustomed to hearing."[100]

By contrast, engineers from the BBC reported that in 1947, after conducting similar tests with the BBC orchestra, all musicians except for the leading trumpet and oboe players had a preference for simple tones.[101] Besides the lack of overall consistency between German and British instrumentalists' reactions, this test also revealed discrepancies within the BBC orchestra members, depending on the instrument played. The engineers and musicians who had run the experiment, Alexander reported, "had been extremely surprised," as had been the members of the BSI pitch committee, who were also "gratified at the result of the B.B.C. tests, which would make the task of presenting the standard note much simpler." The preference of most British musicians for simple tones promised an economic way of implementing the standard in Britain, but also pointed to the diversity in the material embodiments of concert pitch across different national contexts.

CONCLUSION

All in all, the eight years separating the publication of the 1945 *Herald Tribune* article from the 1953 conference in London at once solidified and unsettled the 1939 agreement. Despite the emergence of a strong international consensus about the choice of *A* 440 and the proliferation of new electroacoustic technologies providing unprecedented control over sound, concert pitch continued to escape unification. Postwar negotiations thus ultimately unveiled the limits of the territory within which the ISO would exert its authority. Even as musical standardization was becoming part of the organization's agenda of global governance, it became clear that concert pitch would remain a plural object that could not be entirely reconciled with scientific and industrial standards of uniformity.

Crucially, efforts to implement the 1939 decision reshaped scientific conversations about concert pitch. In turning their attention to musicians, standardizers started to mobilize the conceptual and technological tools of several subfields that had hitherto remained separate from these debates, including psychology and psychoacoustics. This reconfiguration, in turn, enhanced two opposite responses. The first one was an increased control and attempt to normalize not only sound, but also the subjects and objects producing it. Growing evidence of musicians' tendency to divert from any sonic point of reference also fueled opposite claims. Although new acoustic research informed ever more intrusive empirical studies of musicians' behavior, the performers' entrance into the laboratory in turn challenged previous acoustic ideas. This tension revealed wider trends in the history of technical and industrial standards. Some influential voices in the debates cited aesthetic notions of originality and invention to point out pitch's fundamental irreconcilability with industrial and scientific demands for uniformity. Whereas existing works on standards have mostly focused on their producers and production procedures, musical pitch, as a measure that engages the participation of "lay" users at all times, calls for more encompassing approaches, attentive to the social and cultural ramifications of standards and standardization processes.

In 1953, West Germany reported to the ISO on a survey of the opinions of the nation's musical communities about concert pitch. Exhibiting a particular eagerness to cultivate a democratic approach to the question, the authors of this study synthesized their findings by highlighting the plurality of views that their research had uncovered. German musicians, they argued, could be divided into three groups: the first considered that "the musical pitch problem should not be subject to technical considerations and still less to official specifications"; "The selection of the musical pitch should . . . be left entirely to musicians and to orchestras and band leaders." A second opinion was that "While the maintenance of a reproducible musical pitch and the application of electro-acoustic instruments for this purpose are recommended, it is doubtful whether the fixation of a musical pitch a'=440 Hz can be generally maintained." As the authors of the study reported, "The advocates of this opinion suggest to provide instruments permitting the maintenance of adjustable higher musical pitches, such as 442 and 445 Hz." Finally, "A third opinion adheres strictly to the London resolutions and to the corresponding German standardization," especially among "important makers of musical instruments who on account of their adherence to the resolutions have incurred considerable costs in applying them to the construction of their instruments." Drawing on this plurality of views, the authors recommended that

> The resolutions passed in London should be left unchanged if possible. It may, however, prove advisable to leave it to the discretion of the manufacturers of musical instruments to what extent they desire to accept this recommendation . . . Electro-acoustic instruments should be available to interested

> parties desiring to maintain the musical pitch. The question remains as to whether these instruments should be transposable in a reproducible manner from the standard musical pitch to other frequencies.[102]

Perhaps what this proposal revealed most was West German political realities and the authors' reluctance to authoritatively impose any decision over German musical circles. Yet it also reflected a broader tendency in the negotiations to move away from rigid definitions of concert pitch and embrace the plurality that music introduced in this object. While new sonic technologies secured access to ever more accurate embodiments of *A* 440, awareness was also growing of music's irreconcilability with contemporary scientific and industrial standards of precision. On the eve of the ISO conference of 1953, after over a century of research and negotiations, one thing was crystal clear: the territory within which official concert pitch would operate would be limited. Emphasizing the flexibility that characterized theoretical definitions and material embodiments of concert pitch, however, should not overshadow the impact of increasingly invasive technics aimed to discipline sound objects and subjects alike. Further research remains to be done to track the impact of these practices across a wide spectrum of social settings, including professional orchestras, amateur bands, musical instrument factories, conservatories, and schools.

A decade after unprecedented international consensus on concert pitch had been reached in London, the "pitch quarrel" revealed the persistence of considerable divisions about the choice of a suitable standard for music, carrying wider scientific, political, and cultural implications. Although Dussaut's reactionary and nationalistic enterprise may appear an isolated and outdated intervention unworthy of scholarly attention, it requires close examination not only because of its historical significance, but also because it planted seeds that continue to shape debates. Not only is his proposed standard still alive, but since the 1960s, musicians' eagerness to secure the preservation of ancient musical works has resulted in the creation of further alternative measures. The several radically different conceptions of concert pitch did not disappear with the ISO's 1953 validation of the 1939 decision. Rather, the decades since have seen the multiplication of alternative standards, a testament to the survival of musical pitch's plurality.

Epilogue

At the end of this book, the example of our oboist playing her *A* at the beginning of a concert, mentioned in the opening lines of the introduction, resonates differently. This simple, seemingly natural sound encapsulates decades of battles and power relations that renew our understanding of contemporary musical and sonic practices. Concert pitch carries within it music's various political economies, cultural references, as well as the interests and practices of all the groups that sought to tune the world according to their own worldviews. *Tuning the World* shows that connecting the various sites of increasingly interdependent music cultures meant selecting the ones worthy of being the yardsticks of this new global order. The process of fixing a specific frequency for concert pitch was one by which standardizers sought to impose certain musical genres, traditions, styles, time periods, or repertoires as models for the development of contemporary music cultures. In other words, standardization of pitch has always been a way of drawing the contours of what counts as music.

As the foregoing chapters have demonstrated, however, there were important limitations to this program. To begin with, the scope of pitch reformers was narrow. Although experts on pitch envisioned their plans for pitch reforms as global and imagined their standards as universal, in fact, their schemes rarely went beyond Western Europe and the United States. Lissajous and the French government may have mobilized "the globe" as the initial horizon for their reform; what they devised was actually a plan to diffuse the standard within the limited frame of the nation, and then only among musical institutions worthy of the expense or surveillance. Similarly, when broadcasters were walking toward the first solid agreement on an international pitch standard at the end of the 1930s, they deliberately dismissed non-Western contexts as unworthy of their attention. As a piano-maker delegate from Bluthner & Co. eloquently declared: "The discussion that afternoon was really with the idea of arriving at a pitch which could be used at what might be called high-class concerts."[1] As much as standardizers were uninterested in implementing concert pitch beyond elite music circles in Europe and the United States, scholars and musicians outside their world were largely indifferent to the standard. While the diffusion of Western musical instruments, tuning systems, and scales has triggered conversations across a great variety of cultural contexts, including in Asia, the Middle East, and Central America, the creation of a standard pitch appears to have remained under the radar of public attention in these parts of the world until ISO's orchestrated efforts in the aftermath of World War II.

The Western concept of standard pitch and the technologies that embody it were and continue to be unfit to most musical traditions where the voice and a sonically flexible instrumentarium dominate. Transcultural encounters like those initiated by Westerners to document the musics of "others" provide clear evidence of this incongruity. For instance, in 1893 Sir Francis Taylor Piggott, a British jurist who served as adviser to the Japanese prime minister and studied local musical instruments and practices during his years in Japan, reported that

> Although a pitch-pipe is sometimes used, the first string [of the koto], the dominant of the scale is tuned first, and is within limits arbitrary: for a loud singer it is tuned up, for a singer with a small voice it is tuned down. But the normal pitch of the note approximately is middle C; I have, however, taken it as C#. On the Japanese Flute this note lies midway between C and C# on a Piano tuned to Broadwood's Philharmonic pitch.[2]

Fixed tones exist in a wide array of music cultures, but in most cases, they vary in accordance with the circumstances. Furthermore, the introduction of European fixed-pitch instruments into hybrid settings with indigenous instruments in the colonies has triggered strategies aimed at expanding Western definitions of musical pitch, including ornamentation, the application of alternative tuning systems, variations in air pressure, and the introduction of changing-scale technologies.

Critics of Western sonic precision emerged as well in the most elite scientific and musical circles in Europe and the United States. Since the 1930s, musical acousticians have drawn attention to the aesthetic value of discrete pitch variations. They emphasize not only the impossibility, but the undesirability of a total control over sound frequencies. Western art music's valuation of ornamentation techniques such as vibrato attests to the continuities between sonic cultures regardless of the way in which they relate to pitch and tuning standards, while the recent invention of electronic keyboards affording sonic continuity and flexibility further blurs the line between different musical worlds.[3]

The countless problems confronted by standardizers have continued to further shrink the standard's territory of application. In this regard, the situation of postwar European countries did not radically differ from that of mid-nineteenth-century American pioneers of pitch standardization and diplomats of the Royal-Imperial Austro-Hungarian Empire. Whatever musical, economic, or political power these actors may have possessed, it did not grant them sufficient authority to secure sonic uniformity. Surprisingly little progress was ever made in Western powers' quest for sonic control.

All in all, the history of pitch standardization may leave the reader with the impression that she is dealing with the leaky jug of the Danaids. What is it other than a repeated, collective global failure? What can we learn by tracing the laborious efforts of those eager to "tune the world"? The great variety of situations and records of the pitch negotiations in archives and collections across the world offer a unique base through which we may restore complexity to the often overly simplistic narratives that have accompanied the global turn of musicology and sound studies in recent years. In calling for nuanced approaches to sound and power, I mean not merely to diversify the questions that one can ask of such rich material, but to deconstruct the hegemonic gestures that scholars themselves make when, lacking a proper contextualization of actors' discourses, they take the Weberian and Foucauldian notions of rationalization and discipline for granted. Discourses are often performative; and although they do have a significant impact on reality, and undoubtedly play a role in strategies of domination, they also reveal the distress of actors confronted with sound's remarkable tendency for indiscipline.

I have grounded this book in the empirical evidence of pitch negotiation archives not only for the sake of narrative precision, but because the protagonists of the narrative themselves turned to archives of the negotiations to inform their own deliberations. Records of Britain's firm Broadwood, participants in the negotiations from the 1870s to the late 1930s, offer a good example of this phenomenon. The piano builder Evelyn Broadwood brought historical knowledge of previous negotiations, acquired through his research into his company's archives, to the BSI pitch committee during the summer of 1938. Broadwood located Hipkins's 1896 article published in the *Journal of the Royal Society of Arts* (recommending a rise in frequency from the 1859 French pitch),[4] and through his mobilization of this document ensured that the 1939 conference reaffirmed several of the cultural premises and bore many of the marks that had shaped nineteenth-century standardizing efforts.[5]

The constitution of a historical archive of the negotiations triggered a new awareness of pitch's geographic and historical variety, which in turn fueled and continues to fuel both unification and diversification processes across different musical traditions. Since the 1960s, early music performers have been using ever more diverse pitches for their performance of "Baroque," "classical," and "Romantic" repertoires. Nikolaus Harnoncourt was one of the most influential promoters of *A* 415 as an alternative to *A* 440 through his performances with the Concentus Musicus of Vienna, composed of "period" instruments (that is, either originals or copies of originals). During these concerts, the introduction of *A* 415 as the "Baroque pitch" not only "affected the sound quality of both instruments and voices, especially to the ears of people brought hearing only instruments at 440," it also "created a 'fork in the road'" for the careers of Western classical musicians.[6] As the oboist and musicologist Bruce Haynes recalled, the diffusion of *A* 415 "made it virtually impossible to mix Romantic and Baroque instruments in the same ensemble. This in turn forced musicians to choose between the two, defining themselves as 'modern' or 'historical'; there was a symbolic barrier, thrown up by the mundane reality of pitch."[7]

Invoking "authenticity" to justify using a standard for the performance of musical works dating from an era characterized by the lack of such a unifying point of reference is something of a paradox. Over the last thirty years, however, musicologists have expanded the spectrum of standard pitches available for classical music performances by drawing on, and deepening, nineteenth-century inquiries into the historical variations of pitch. Musicians have used this research to provide audiences with more "authentic"—or simply sonically more varied—interpretations of early music. It has now become customary to play French seventeenth- and eighteenth-century operas at *A* 392 and most of Bach's and Handel's works at *A* 421, while using *A* 430 for Mozart, Haydn, and other "Romantic" composers, *A* 460 for Bach's works before his Leipzig period, and *A* 465 for seventeenth-century Venetian compositions. More recently, the conductor Jérémie Rhohrer has introduced *A* 432 as "Verdi's" pitch, following a complicated history of claims and conspiracies that have surrounded this standard since Dussaut's campaign in the late 1940s and 1950s.[8] Although this diversification in standard pitches seems to undermine the project of unifying musical frequencies, it is in fact a prolongation of the standardization process: only against the backdrop of a standardized musical world do other tones take on a different color; the knowledge that musicians use to tune their instruments was a product of the negotiations.

"Historical" performers are not the only ones to both draw on and take their distance from concert pitch *A* 440. If the oboist I mentioned in the introduction was a member of one of today's leading orchestras worldwide, it is likely that she would in fact play at *A* 444 (were she a member of the Gewandhaus in Vienna or of the Boston Symphony Orchestra), *A* 443 if she performed within the Berlin Philharmoniker, or *A* 442 if she was part of the Chicago Symphony Orchestra. These strategies of diversification all attest to the resilience of one of this book's central themes: the shared belief, since the beginning of the negotiations, that "the higher the pitch, the brighter the sound." Although further work remains to be done to document more finely performers' approaches to the standard, it is clear that fixing pitches in Western music has opened the door to ever more diversification in musicians' use of frequencies. The documentary *Pianomania*, for example, nicely captured the central role played today by tuners in solo pianists' constitution of their idiosyncratic sound. Tuning, in other words, has become an essential part of the aural aesthetic.

These phenomena are not specific to classical music. In the field of popular music, too, the standardization of musical sounds has resulted in a similar dialectic between standardization and diversification. Despite the massive use of acoustic standards in this field, the spread of tuning devices and standard electronic instruments (keyboards in particular) has resulted in a quest by musicians for sonic precision that is also a search for singularity. The proliferation of digital applications offering dozens of tuning systems intended for popular music ensembles and the diversification of scales reveal the same complex of pitch standardization, acoustic knowledge, and aesthetic individuation. What is more, a growing number of blogs accuse *A* 440 of distorting music and of producing harmful effects on the body. They promote instead the use of *A* 432, presenting it as a natural, immemorial standard.[9]

In addition to performers, pitch standardization has shaped the practices of composers up to this day. Since the Second World War, the fixation of musical frequencies has triggered the development of microtonal works and theories that decompose sound into infinitesimal units, the invention of ad hoc instruments, experiments around the tuning of orchestral instruments, as well as the use of tuning forks and other scientific instruments as musical elements. More specifically, the adoption of concert pitch has elicited responses ranging from humor to resistance and poetry. In 1947, Edgar Varèse left unfinished a piece entitled *Tuning Up*, which playfully stages the ritual of orchestras' tuning. The following year, Pierre Schaeffer composed a *Diapason Concertino* that similarly refers humorously to the recently adopted frequency *A* 440. More recently, the composer Mark-Anthony Turnage has used differences in pitch in his work *About Time* to stage the divisions between the intervals that continue to govern performance practice. More fundamentally, composers' search for new systems of tonic organizations attest to their desire to emancipate themselves from the standardization of frequencies.

The dialectic between unification and diversification also infuses the work of several sound artists who have dived into archives of the debates. For example, Ryoji Ikeda's ongoing project *A* uses data produced by sound standardizers to undermine their very project by reviving the historical and geographic diversity of frequencies. He turns nineteenth-century acousticians' observations of beats and sound interferences—the point of departure for the measurement and control of pitch—into a poetic and liberating experience.[10] Other artists evince the permeability of their work and media archeology. Richard Chartier, for instance, drew on his exploration of Koenig's great tonometer—a set of 670 tuning forks covering four octaves—at the Smithsonian's National Museum of American History for his *Transparency (Performance)* (2011), and Nicolas Bernier's *Frequencies* series is anchored in his discovery of the University of Rennes's (France) collection of nineteenth-century tuning forks.[11] As rich as the archival and material traces left by the negotiations are for historical, sonic, and musical explorations, they nevertheless have important limitations. These collections have many blind spots, and relying on them undoubtedly leads to an overrepresentation of the voices and actions of the archives' creators. Future research may try to map out what's missing, what these silences have to tell us, and where they can lead us, both as historians and as creators.

Further historical, anthropological, and ethnographic inquiries will undoubtedly make clear just how dual the role of a standard pitch has been, and still is, across the world's music cultures. While primarily intended to connect different eras and areas of music making, a fixed pitch also separates, hierarchizes, otherizes in ways that are all the more insidious in that they have become inaudible, or at least disconnected from their historical, political, epistemological, or industrial resonances. As we come to the end of this book, it seems clear that efforts to play *with* others, be they chronologically or spatially remote, have always also meant playing *without*, or even *against* further historical or geographical others. If we want to trace the ways in which musical practice shapes culture and politics, we thus need to start paying more attention to how sound, even in its seemingly most inoffensive, "natural" version, is always loaded with values that merit our scholarly investigation.

Acknowledgments

This book has benefited from the help and support of many. There are no words to express how fortunate I have felt from the beginning to the end of this project. When I started researching the history of concert pitch in 2016, I had no idea where this work would take me. I especially did not suspect how much it would change my identity as a scholar. After a dissertation on the history of concert life in nineteenth-century French churches, which had allowed me to engage with cultural and religious historians, mainly in France, this project has led me to many new disciplinary and academic territories—from history of science and science and technology studies to media and sound studies, to global, postcolonial, and transnational history, in several countries. Most importantly, writing this book has enabled me to meet numerous brilliant colleagues, several of whom have become regular collaborators and friends. Writing this book has thus reshaped my career and life in the most unexpected and fulfilling ways.

I warmly thank Marta Tonegutti, James Davies, and Nicholas Mathew for their wonderful insights on the manuscript, and for helping me navigate the review process so smoothly. Paul de Angelis has provided unique guidance for the revisions and made the book far more readable. Edward Gillin, Annelies Fryberger, Emmanuel Reibel, Ben Steege, Alix Hui, and an anonymous reviewer have generously engaged with the manuscript, and their clever comments have helped me explore many unforeseen questions and considerably sharpen my arguments: many thanks to them. Marianne Tatom and Stephen Twilley were great helps in editing the manuscript.

All my gratitude goes to Viktoria Tkaczyk, who welcomed me twice in her research group "Epistemes of Modern Acoustics" at the Max Planck Institute for the History of Science over the course of my research: first for a month, as I was starting archival work, in the fall of 2016, then for two years, as a research scholar, in 2018–2019. These stays have shaped my research in many ways—and will probably remain some of my happiest professional experiences. They gave me the opportunity to visit several archives and collections in Europe and the United States, to host workshops and conferences, and to invite colleagues to collaborate on writing projects. Throughout the months I spent at the MPI, I was surrounded by a vibrant interdisciplinary community of scholars: my deepest thanks to Kate Sturge, Joeri Bruyninckx, Xiaochang Li, Hannah Wiemer, Lotte Schüßler, Leendert van der Miesen, João Romão, and Leon Chisholm, as well as Andrea Bohlman, Karin Bijsterveld, David Pantalony, Jonathan Sterne, Veit Erlman, Martin Scherzinger, and many others. Birgitta von Mallinckrodt expertly and warmly handled the organization of research trips and guests' visits, leaving us with nothing else to do but enjoy our work and her delicious cakes. The group's research assistants, Jonathan Haid, Hannah Eßler, Yaroslav Koshelev, Julia Steinmetz, Alina Topf, and Fabian Voigtschild, have been the best possible helps. Myles Jackson introduced me to this wonderful community, after generously agreeing to meet with me in New York in February 2016, as I was just starting this project: I can't thank him enough for this.

In its early stages, this book received the support of the Fulbright Commission, which funded my stay at UCLA in 2016–2017, where I had the chance to discuss this project with members of the musicology and the history of science departments: Nina Eidsheim, Robert Fink, Elisabeth LeGuin, Raymond Knapp, Tamara Lewitz, Mitchell Morris, Shana Redmond, and Jessica Schwartz for the former; Amir Alexander, Soraya de Chadarevian, Iris Clever, Joshua McGuffie, Ted Porter, Mary Terrall, and Norton Wise for the latter, who all welcomed me like one of them and gave me a speedy and exciting training in history of science. This project was also funded by the Huntington Library, where I had the privilege and joy to discuss my research in the most beautiful environment with brilliant scholars, especially Andrew Lipman and Sophie Maisonneuve. Suzannah Clark, Katharine Ellis, and Patrice Veit supported every single fellowship application I submitted to fund this research; I can't thank them enough for their generosity and unfailing support.

Since 2019, this project has received the support of the STMS Lab at IRCAM (Centre national de la recherche scientifique-IRCAM-Sorbonne Université). I am grateful to Brigitte d'Andréa-Novel and Jean-Louis Giavitto for welcoming me there, as well as to the members of the research team

"Analyse des Pratiques Musicales"—Nicolas Donin, Clément Canonne, François-Xavier Féron, and Laurent Feneyrou—for all their feedback on the project, and for being wonderful colleagues. I also thank the participants of our reading seminar who have commented on chapters of the book: Isabelle Mayaud, Gesa zur Nieden, Jonathan Goldman, and Benoit Haug. Anne-Marie Vaudeville and Viktoriya Uscumlic have been wonderful supports with administrative and financial matters—many thanks to them too. Several other colleagues in France helped shape the project at its different stages: Andreas Mayer, Anne Rasmussen, François Picard, Florence Gétreau, and Claire Lemercier. I also thank my colleagues from New York University's Music Department—Michael Beckerman, Brigid Cohen, Christine Dang, Martin Daughtry, Elizabeth Hoffman, Louis Karchin, Maureen Mahon, Jaime Oliver, David Samuels, and Suzanne Cusick—for their confidence, stimulating feedback, and inspiring example as a bright, creative, and happy scholarly community. I look forward to pursuing my research as one of them in the years to come. Thanks to Cayla Delardi and Freeman Williams for handling administrative questions so efficiently.

Many archivists and heads of collections have generously shared their knowledge and given me exciting tours of their holdings. In particular, I thank Mr. Klein, who helped me navigate the Bundesarchiv in Berlin; Steven Turner for showing me the National Museum of American History's acoustic collections and discussing my project; Gérard Emptoz, Daniel Blouin, and the recently departed Denis Beaudoin for opening to me the doors of the Société d'Encouragement pour l'Industrie Nationale; Franz Sachslehner for showing me the archives and collections of the University of Vienna; Giovanni Organtini for giving me a tour of the Museum of Physics at the University La Sapienza; and William Fickinger for welcoming me so warmly at Case Western and sharing his knowledge of William Miller's work.

Partial results of this research have been published in articles, often in special issues and edited volumes. Thanks to their editors Frédéric Ramel, Cécile Prévost-Thomas, Joëlle Le Marec, François Ribac, Christophe Levaux, Antoine Hennion, Hansjakob Ziemer, Julia Kursell, and Kirsten Paige for inviting me to participate in their exciting projects and for all their feedback. I am also grateful to the anonymous reviewers of these articles, who helped me think through this project. Thanks also to Rebekah Arendt, Damien Mahiet, and Andrew Hicks for inviting me to present my work on the panel "The Politics of Harmony" they organized at AMS in 2019, and to the numerous colleagues who have provided me with references, feedback, and thoughts at each of the seminars and conferences where I have presented this research.

The idea of this project was born during a Sunday walk with my husband, Ian, at Les Buttes-Chaumont sometime in 2015 and has bloomed over the course of our innumerable conversations. This book is a testimony to the creativity and energy I get from our relationship. As I am writing these acknowledgments (Paris, November 2021), we are getting ready to welcome our daughter into this world. The thought of her birth has driven the last months of work on this project in the most exciting ways.

Notes

INTRODUCTION

"The Review's Technical Department: On a Uniform Pitch," *The Music Trade Review* 51, no 9 (27 Aug. 1910): 17.

1 Marin Mersenne, *Harmonie universelle, contenant la théorie et la pratique de la musique*, 3 vols. (Paris: Sebastien Cramoisy; reprint ed. Paris: Centre national de la recherche scientifique, 1965). *Livre Troisième des instruments à chordes*, Proposition 23, 147–151; Michael Praetorius, *Syntagma Musicum II. De Organographia*, trans. David Z. Crookes (Oxford: Clarendon Press, 1986 [1618–1620 for the 1st ed.]), part II, chapter 2, 30–33.

2 See below, chapter 1.

3 It was first adopted as ISO Recommendation R16, in 1955, then as ISO 16, in 1975 (https://www.iso.org/standard/3601.html). See below, chapter 5.

4 For example, in 2018, the Théâtre des Champs-Élysées in Paris hosted performances of Verdi's *Traviata* using *A* 432.

5 Ariejan Korteweg, "Vrijzinnige Partij: verlaag de grondtoon, deze wekt verdeeldheid en agressie op," *Volkskrant*, 3 March 2017, https://www.volkskrant.nl/nieuws-achtergrond/vrijzinnige-partij-verlaag-de-grondtoon-deze-wekt-verdeeldheid-en-agressie-op~b718e98c/. See Fanny Gribenski, "Tuning Forks as Time Travel Machines: Pitch Standardization and Historicism," in V. Tkaczyk and L. van der Miesen (eds.), *Sonic Things: Knowledge Formation in Flux. Sound Studies* 6 (2020/2): 153–173.

6 I borrow the term "impossible" in this section's title from Chandra Mukerji, *Impossible Engineering: Technology and Territoriality on the Canal du Midi* (Princeton, NJ: Princeton University Press, 2009). Far from the successful "distributed reasoning" and visible triumph of Mukerji's actors, however, the history of concert pitch is one of resounding failures.

7 Émile Leipp and Michèle Castellengo, "Du diapason et de sa relativité," *Revue musicale* 294 (1977): 7–10; and Bruce Haynes, *A History of Performing Pitch: The Story of "A"* (Lanham, MD: Scarecrow Press, 2002).

8 "Le diapason ayant monté d'un ton en cent ans ou d'un demi-ton en un demi-siècle, si sa marche ascendante continuait, il parcourrait en six cents ans les douze demi-tons de la gamme, et serait nécessairement en l'an 2458 haussé d'une octave." Hector Berlioz, "Le Diapason," *Journal des débats* (29 Sept. 1858): 2. Unless otherwise indicated, all translations from French are my own.

9 William Weber, *The Rise of Musical Classics in Eighteenth-Century England: A Study in Canon, Ritual and Ideology* (Oxford: Clarendon, 1992); and "La musique ancienne in the Waning of Ancien Régime," *Journal of Modern History* 56, no. 1 (March 1984): 58–88.

10 Katharine Ellis, *Interpreting the Musical Past: Early Music in Nineteenth-Century France* (Oxford: Oxford University Press, 2000); and James Garrat, *Palestrina and the German Romantic Tradition. Interpreting Historicism in Nineteenth-Century Music* (Cambridge: Cambridge University Press, 2010).

11 "Resolution (71) 16 (Adopted by the Ministers' Deputies on 30 June 1971). The Standardisation of the Initial Tuning Frequency."

12 Emily Dolan, *The Orchestral Revolution: Haydn and the Technologies of Timbre* (Cambridge: Cambridge University Press, 2013).

13 Hector Berlioz, "Revue musicale. Théâtre de l'Opéra-Comique," *Journal des débats* (24 Sept. 1856): 2.

14 "Chronique musicale. Sur le Diapason de l'Académie Royale de Musique," *Journal des débats* (14 April 1824): 4.

15 In 1945, reacting to France's adoption of the American standard *A* 440 instead of its diapason normal, the music critic René Dumesnil wrote: "If we continued like this, jazz would forever replace classical music." [Si l'on continuait ainsi, le jazz remplacerait définitivement la musique classique.] "La Question du diapason," *Le Monde* (15 Sept. 1945).

16 Albert de Lassalle, "Chronique musicale. La question du *diapason* et sa prochaine solution officielle," *Le Monde illustré* (20 Nov. 1858): 335.

17 CGPM, *Compte rendu des séances de la 11e conférence générale des poids et mesures* (Paris: Gauthier-Villars, 1961).

18 Pythagorean tuning was the first system of tonic organization to be discussed in Western medieval theory. Initially described during Greek Antiquity, it derives intervals from natural fifths—the interval between the second and third tones of the overtone series. This system produces sharp thirds, which became increasingly problematic by the end of the medieval period, with the unprecedented development of polyphony. Theorists addressed this issue through the creation of various "meantone temperaments," which prioritized consonant vertical thirds suited to polyphonic music over perfect fifths. Zarlino (1517–1590) was the first theorist to describe an alternative system that gained a lot of authority: just intonation. In this system, intervals are derived from natural fifths and natural thirds. See Ross W. Duffin, *How Equal Temperament Ruined Harmony (and Why You Should Care)* (New York: W. W. Norton, 2007).

19 See below, chapter 1.

20 *Le Diapason et la notation musicale simplifiés* (Paris: Schott, 1873), 14.

21 See Fanny Gribenski, "Words and Numbers: The Many Languages of Nineteenth-Century Pitch Standardization," in J. Kursell, V. Tkaczyk, and H. Ziemer (eds.), *Sounds of Language, Languages of Sound: Themes and Tools in the Humanities. History of Humanities* 6, no. 1 (2021): 11–34.

22 It was considered to be Paris's average temperature and often associated with scientific experiments. It appears in the catalogue of the acclaimed maker Joseph Marloye, who coined the term "diapason normal" in 1840, in reference to a device giving a *C* producing 512 vibrations commonly adopted in physics laboratories. *Supplément au Catalogue de Marloye* (Paris: Ducessois, 1840), n.n.

23 See below, chapter 5.

24 Stefan Timmermans and Steven Epstein, "A World of Standards but Not a Standard World: Toward a Sociology of Standards and Standardization," *Annual Review of Sociology* 36, no. 2 (2010): 71.

25 Simon Schaffer, "La nouvelle definition du kilo est une révolution théologique," Le Monde (13 Nov. 2018).

26 For studies of standards, see, in particular: M. Norton Wise and Crosbie Smith, "Measurement, Work and Industry in Lord Kelvin's Britain," *Historical Studies in the Physical and Biological Sciences* 17, no. 1 (1986): 147–173; Bruce J. Hunt, "The Ohm Is Where the Art Is: British Telegraph Engineers and the Development of Electrical Standards," *Osiris* 9, "Instruments" (1994): 48–63; Heinz Otto Sibum, "Reworking the Mechanical Value of Heat: Instruments of Precision and Gestures of Accuracy in Early Victorian England," *Studies in the History and Philosophy of Science* 26, no. 1 (1995): 73–106; M. Norton Wise, ed., *The Values of Precision* (Princeton, NJ: Princeton University Press, 1995); Graeme J. N. Gooday, *The Morals of Measurement: Accuracy, Irony, and Trust in Late Victorian Electrical Practice* (Cambridge: Cambridge University Press, 2004); Theodore M. Porter, *Trust in Numbers: The Pursuit of Objectivity in Science and Public Life* (Princeton, NJ: Princeton University Press, 1996); Simon Schaffer, "Physics Laboratories and the Victorian Country House," in *Making Space for Science: Territorial Themes in the Shaping of Knowledge*, ed. Crosbie Smith and Jon Agar (Basingstoke: Palgrave, 1998), 149–180; and Simon Schaffer, "Late Victorian Metrology and Its Instrumentation: A Manufactory of Ohms," in *The Science Studies Reader*, ed. Mario Biagioli (New York: Routledge, 1999), 457–478.

27 See, for instance: JoAnn Yates and Craig N. Murphy, *Engineering Rules: Global Standard Setting since 1880* (Baltimore, MA: Johns Hopkins University Press, 2019); Martha Lampland and Susan Leigh Starr, *Standards and Their Stories: How Quantifying, Classifying, and Formalizing Practices Shape Everyday Life* (Ithaca, NY: Cornell University Press, 2009); and Nils Brunsson and Beigt Jacobson (eds.), *A World of Standards* (Oxford: Oxford University Press, 2000).

28 Historian of science Myles Jackson has shown that the relationship between music and science in nineteenth-century Germany materialized through musical standardization. *Harmonious Triads: Physicists, Musicians, and Instrument Makers in Nineteenth-Century Germany* (Cambridge, MA: MIT Press, 2006). Historians of media and technology Emily Thompson, Karin Bijsterveld, Mara Mills, and Jonathan Sterne have analyzed the ways in which standardization processes have shaped the various facets of our modern soundscapes, from noise management to hearing tests, and from car design to the creation of formats intended to facilitate

the circulation of recorded music. Emily Thompson, *The Soundscape of Modernity: Architectural Acoustics and the Culture of Listening in America, 1900–1933* (Cambridge, MA: MIT Press, 2002); Karin Bijsterveld, *Mechanical Sound: Technology, Culture and Public Problems of Noise in the Twentieth Century* (Cambridge, MA: MIT Press, 2008); Karin Bijsterveld, Eefje Cleophas, Stefan Krebs, and Gijs Mom, *Sound and Safe: A History of Listening Behind the Wheel* (Oxford: Oxford University Press, 2013); Mara Mills, "Deafening: Noise and the Engineering of Communication in the Telephone System," *Grey Room* 43 (Spring 2011): 118–143; Mills, "Decibel," in *Handbuch Sound. Geschichte—Begriffe—Ansätze*, ed. Daniel Morat and Hansjakob Ziemer (Stuttgart: J. B. Metzler Verlag, 2018), 52–56; and Jonathan Sterne, *MP3: The Meaning of a Format* (Durham, NC: Duke University Press, 2012).

29 See Roger Bowers, "The Performing Pitch of English 15th Century Church Polyphony," *Early Music* 8, no. 1 (1980): 21–28; Barbara Owen, "Pitch and Tuning in Eighteenth and Nineteenth Century American Organs," *The Organ Yearbook* 15 (1984): 54–59; Arthur Mendel, "On the Pitches in Use in Bach's Time," *Musical Quarterly* 41 (1995): 332–354 and 466–480; and Haynes, *A History of Performing Pitch*.

30 Viktoria Tkaczyk, *Thinking with Sound: A New Program in the Sciences and Humanities around 1900* (Chicago: University of Chicago Press, 2022).

31 Carmel Raz, "Of Sound Minds and Tuning Forks: Charcot's Acoustic Experiments at the Salpêtrière," *Musicology Now* (22 Oct. 2015), https://musicologynow.org/of-sound-minds-and-tuning-forks-charcots-acoustic-experiments-at-the-salpetriere/; Mills, "Deafening."

32 See especially Jackson, *Harmonious Triads*; Alexandra Hui, *The Psychophysical Ear: Musical Experimental Sounds, 1840–1910* (Cambridge, MA: MIT Press, 2013); Peter Pesic, *Music and the Making of Modern Science* (Cambridge, MA: MIT Press, 2014); Alexandra Hui, Julia Kursell, and Myles W. Jackson, "Music, Sound, and the Laboratory," *Osiris* 28 (2013); Benjamin Steege, *Helmholtz and the Modern Listener* (Cambridge: Cambridge University Press, 2012); and James Q. Davies and Ellen Lockhart (eds.), *Sound Knowledge: Music and Science in London, 1789–1851* (Chicago: University of Chicago Press, 2016).

33 For a general approach to the question of boundary making in relation to science in the nineteenth century, see Thomas F. Gieryn, "Boundary-Work and the Demarcation of Science from Non-Science: Strains and Interests in Professional Ideologies of Scientists," *American Sociological Review* 48, no. 6 (Dec. 1983): 781–795. On the relation between music, science, and instrument making, see Jackson, *Harmonious Triads*; and Hui, Kursell, and Jackson, "Music, Sound, and the Laboratory."

34 I use this historically contested term to designate a process that started after 1760 in Britain.

35 In 1815, following several natural philosophers' attempts to improve Étienne Loulié's chronometer, the German inventor of musical instruments and automata patented the metronome, which was actually created by the Dutch mechanician Dietrich Nikolaus Winckel. See Jackson, *Harmonious Triads*, 185–197; and Aurélie Barbuscia, "La pratique musicale, entre l'art et la mécanique. Les effets du métronome sur le champ musical au XIXe siècle," *Revue d'histoire du XIXe siècle* 45, no. 2 (2012): 53–68.

36 On new organology, see Emily Dolan, *The Orchestral Revolution*; and from the same and John Tresch, "Toward a New Organology: Instruments of Music and Science," *Osiris* 28, no. 1 (2013): 278–298. On embodiment, see especially Elisabeth Le Guin, *Boccherini's Body: An Essay in Carnal Musicology* (Berkeley: University of California Press, 2006); and Holly Watkins and Melina Esse, "Down with Disembodiment; or, Musicology and the Material Turn," *Women and Music: A Journal of Gender and Culture* 19, no. 2 (2015): 160–168. For particularly influential examples of science and technology studies approaches to sound and music, see Georgina Born, "Music and the Materialization of Identities," *Journal of Material Culture* 16, no. 4 (2011): 376–388; and Eliot Bates, "The Social Life of Instruments," *Ethnomusicology* 56, no. 3 (2012): 363–395.

37 In recent years, however, music scholars' turn from technologies to infrastructures and supply chains has offered a sociopolitical and ecological critique. See especially Kyle Devine, *Decomposed: The Political Ecology of Music* (Cambridge, MA: MIT Press, 2019); and by the same and Alexandrine Boudreault-Fournier (eds.), *Audible Infrastructures: Critical Conjunctures in Music and Sound* (Oxford: Oxford University Press, 2021).

38 Gooday, *The Morals of Measurement*; and Porter, *Trust in Numbers*.

39 Ken Alder, "A Revolution to Measure: The Political Economy of the Metric System in France," in Wise, *The Values of Precision*.

40 Vanessa Ogle, *The Global Transformation of Time: 1870–1950* (Cambridge, MA: Harvard University Press, 2015).

41 Mersenne, *Harmonie universelle. Livre Troisième des instruments à chordes*, Proposition 18, 147.

42 Raymond Murray Shafer, *Soundscape: Our Sonic Environment and the Tuning of the World* (New York: Alfred A. Knopf, 1977).

43 Germany is at the center of Jackson's, Steege's, and Hui's works; Sterne, Thompson, and Mills have focused on the United States; Lockhart and Davies on Britain. Bijsterveld's *Mechanical Sounds* embraces a broader geographic scope, but does not really explore transnational connections between the different contexts she considers.

44 Ronald Radano and Tejumola Olaniyan, eds., *Audible Empire: Music, Global Politics, Critique* (Durham, NC: Duke University Press, 2016); Gavin Steingo and Jim Sykes (eds.), *Remapping Sound Studies* (Durham, NC: Duke University Press, 2019); and Peter McMurray and Priyasha Mukhopadhyay (eds.), *Acoustics of Empire* (forthcoming).

45 On these genealogies, see Gribenski, "The Many Languages of Nineteenth-Century Pitch Standardization."

CHAPTER ONE

1 "Plus le ton sera élevé, plus le son sera brillant." *Rapport et Arrêtés pour l'établissement en France d'un diapason musical uniforme* (Paris: Imprimerie impériale, 1859).

2 "Je veux l'émotion coûte que coûte, sans souci des cordes vocales de mon prochain." Charles Jacob Marchal de Calvi, "Libre remontrance au sujet du diapason France," *France musicale* 23, no. 10 (6 March 1859): 106.

3 "Les conditions, et notamment la condition de tonalité, peuvent-elles, doivent-elles être les mêmes, je le demande, suivant qu'il s'agit de la musique vocale, de la musique instrumentale ou de la musique voco-instrumentale?" Marchal, "Libre remontrance au sujet du diapason," 105.

4 "Comment! vous soumettrez à la même loi la corde vocale et la corde de violon, la muqueuse et le cuivre!" Marchal, "Libre remontrance au sujet du diapason," 106.

5 Jules-Antoine Lissajous, "Note sur l'élévation progressive du diapason des orchestres depuis Louis XIV jusqu'à nos jours et sur la nécessité d'adopter un diapason normal et universel," *Bulletin de la Société d'Encouragement pour l'Industrie Nationale* 54, no. 2 (1855): 297.

6 Lissajous defended a dissertation in physics in 1850 entitled "Sur la position des noeuds dans les lames qui vibrent transversalement" ("On the position of nods in transversally vibrating blades"). After three unsuccessful applications to the Académie des sciences, he became a corresponding member of the institution just one year before his death, in 1880. Besides his efforts to integrate the institution, Lissajous was very active as an expert in acoustics among Parisian scientific and industrial circles. In 1857, John Tyndall invited him to present his curves in front of the Royal Institution in London. Serge Benoit, Daniel Blouin, Jean-Yves Dupont, and Gérard Emptoz, "Chronique d'une invention: Le phonautographe d'Édouard-Léon Scott de Martinville (1817–1879) et les cercles parisiens de la science et de la technique," *Documents pour l'histoire des techniques* 17, no. 1 (2009) (Appendix: "Jules-Antoine Lissajous [1822–1880], un physicien engagé dans la valorisation de la recherche").

7 Chladni published the first extensive German treatise on sound, *Die Akustik*, in 1802. Following this publication, he embarked on a Continental tour to advertise his work. It was in Paris that Chladni provoked the greatest interest, securing an audience at the Imperial Palace of the Tuileries with Napoleon, on 7 February 1809. There, before an august gathering of scientific and musical authorities, Chladni performed his experiments with vibrating plates and sand. Viktoria Tkaczyk, "The Making of Acoustics around 1800, or How to Do Science with Words," in *Performing Knowledge 1750–1850*, ed. Mary Helen Dupree and Sean B. Franzel (Boston: De Gruyter, 2015), 27–55.

8 "un spectacle curieux pour les savants comme pour les gens du monde"; "l'exaltation musicale et fébrile de Quasimodo dans la Notre-Dame de Paris, s'enivrant des sons éclatants, étourissants de sa chère cloche." Henri Blanchard, "Du diapason harmonique," *Revue et Gazette musicale de Paris* 12 (3 Aug. 1845): 252–253. On this demonstration, see also Rodolphe Radau, *L'Acoustique ou les phénomènes du son* (Paris: Hachette, 1867), 200–201.

9 Gabriel Finkelstein, "M. du Bois-Reymond Goes to Paris," *British Journal for the History of Science* 36, no. 3 (2003): 261–300. On nineteenth-century scientific performances, see also Bernard Lightman, "The Visual Theology of Victorians' Popularizers of Science," *Isis* 91, no. 4 (Dec. 2000): 651–680.

10 On the Société d'encouragement pour l'industrie nationale (hereafter SEIN), see Serge Benoit, Gérard Emptoz, and Denis Woronoff (eds.), *Encourager l'innovation en France et en Europe. Autour du bicentenaire de la Société d'encouragement pour l'industrie nationale* (Paris: Comité des travaux historiques et scientifiq-France, 2006).

11 "La France possède aujourd'hui une collection complète et authentique des diverses mesures. Les soins apportés dans la confrontation des étalons secon-

daires avec les prototypes déposés aux archives, les moyens employés pour contrôler, sans cesse, l'exactitude des mesures commerciales et industrielles, assurent la conservation indéfinie de cet admirable système. Il serait à désirer que les mêmes principes fussent appliqués à l'établissement et au maintien du diapason qui sert, en quelques sorte, d'unité sonore et dont il n'existe aujourd'hui aucun étalon officiel." Lissajous, "Note sur l'élévation progressive du diapason," 293.

12 Calculated between 1792 and 1799, the meter was claimed to be one ten-millionth of a quadrant of Earth's circumference.

13 Lissajous, "Note sur l'élévation progressive du diapason des orchestres," 297; on the metric system, see Alder, "A Revolution to Measure."

14 In 1834, for example, the SEIN gave a bronze medal to Aristide Cavaillé-Coll, a young prodigy and future leader in the field of organ building who later became a member of the society. Serge Blouin, "Progrès technique et développement musical: Le rôle de la Société d'Encouragement pour l'Industrie Nationale aux XIXe et XXe siècles," https://drive.google.com/file/d/0B8JIO93c80DXTlI2YzA4clBnLXM/edit?resourcekey=0-1QWTHU0UR1qnzNMvJnxitw.

15 Blouin, "Progrès technique et développement musical"; Benoit et al., "Chronique d'une invention: le phonautographe."

16 Lissajous, "Note sur l'élévation progressive du diapason," 294.

17 "Que de belles voix brisées avant d'aborder la scène, et combien peu résistent aux exigences du théâtre! Que de chanteurs éminents passent une moitié de leur vie d'artiste à ruiner les moyens que la nature leur a donnés, et emploient l'autre moitié à dissimuler, à force d'art, les ruines prématurées d'un organe qui fait défaut à leur talent. Telle est la conséquence fatale de l'ascension du diapason." Lissajous, "Note sur l'élévation progressive du diapason," 295.

18 "la méthode employée vulgairement pour régler les diapasons les uns sur les autres . . . à l'aide de la lime." "Au moment où [le diapason] vient d'être réglé, il est d'accord avec le diapason primitif; mais quand il sera refroidi, il montera." Lissajous, "Note sur l'élévation progressive du diapason," 296.

19 "As we deal with this determination, I shall offer a new method based on the drawing of vibrations by the diapason itself." "Je me réserve, lorsqu'on s'occupera de cette détermination, de proposer une méthode nouvelle, fondée sur le tracé des vibrations par le diapason lui-même." Lissajous, "Note sur l'élévation progressive du diapason," 297n3.

20 Lissajous, "Note sur un moyen nouveau de mettre en évidence le mouvement vibratoire des corps," *Comptes rendus hebdomadaires des séances de l'Académie des sciences* 41 (16 July 1855): 93–94; "Note sur une méthode nouvelle applicable à l'étude des mouvements vibratoires," *Comptes rendus hebdomadaires des séances de l'Académie des sciences* 41 (12 Nov. 1855): 814–817. These experiments were subsequently submitted as a *Mémoire sur l'étude optique des mouvements vibratoires présenté à l'Académie des sciences* (Académie des sciences: Dossier Lissajous) in 1857, as well as within a printed and slightly different version of the text published the same year: "Mémoire sur l'étude optique des mouvements vibratoires," *Annales de chimie et de physique* 51 (1857): 147–231. They were also part of Lissajous's *Étude optique des sons* (Paris: Ch. Lahure, 1864). On sound visualization, see Olivier Darrigol, "The Analogy between Light and Sound from Malebranche to Thomas Young," *Physic: Rivista internazionale di storia della scienza* 46, no. 1–2 (2009): 111–217; by the same, "The Analogy between Light

and Sound in the History of Optics from the Ancient Greeks to Isaac Newton (Part II)," *Centaurus* 52, no. 3 (2010): 206–257; and Steven Turner, "Demonstrating Harmony: Some of the Many Devices Used to Produce Lissajous Curves before the Oscilloscope," *Rittenhouse* 11, no. 2 (1997): 33–51.

21 "étudier, sans le secours de l'oreille, toute espèce de mouvements vibratoires, et par suite toute espèce de sons." Lissajous, "Mémoire sur l'étude optique des mouvements vibratoires," 147.

22 "Je colle sur la face convexe du diapason, à l'extrémité d'une des branches, une petite plaque polie qui fait l'office de miroir; je regarde dans ce miroir l'image réfléchie d'une bougie placée à quelques mètres de distance, puis je fais vibrer le diapason; je vois aussitôt l'image s'élargir dans le sens de la longueur des branches." Lissajous, "Mémoire sur l'étude optique des mouvements vibratoires," 147.

23 "Les mouvements vibratoires qui déterminent la production d'un son, s'effectuent avec une telle rapidité que [l'on] n'a pas le temps de les saisir." Lissajous, "Note sur un moyen nouveau de mettre en évidence le mouvement vibratoire des corps," 93. The fast movement of sonorous vibrations and their means of observation had been a problem since early modern sound quantification. Experimenting with the monochord, for instance, the scholar Marin Mersenne (1588–1648) suggested the use of a very long string in order to be able to visibly observe and count its movements. Mersenne, *Harmonie universelle. Livre Troisième des instruments à chordes*, Proposition 23, 147. Lissajous's method drew more specifically from the experiments of Charles Wheatstone, the British natural philosopher and instrument maker who had taken advantage of the principle of retinal persistence to demonstrate the phenomenon of vibration.

24 Paul Quentin Desains, *Leçons de physique*, vol. 2 (Paris: Dezobry, E. Magdeleine et Ce, 1860).

25 "à tout un auditoire nombreux," "à grande distance." These phrases only appear in the manuscript version of the *Mémoire*. In the printed text, Lissajous simply writes "to a whole audience" ("à tout un auditoire").

26 Louis Figuier, *Les Merveilles de la science ou description populaire des inventions modernes*, vol. 2 (Paris: Furne, Jouvet et Cie, 1868), 387.

27 Mersenne, *Harmonie universelle, Livre troisième*, 150.

28 "Toute espèce de corps vibrant." Lissajous, "Note sur un moyen nouveau de mettre en évidence le mouvement vibratoire des corps," 93.

29 Lissajous, "Note sur un appareil simple qui permet de constater l'interférence des ondes sonores," *Comptes rendus hebdomadaires des séances de l'Académie des sciences* 40 (15 Feb. 1855): 133–135.

30 "De tous les corps que l'on peut faire vibrer le plus commode à mettre en experience." Lissajous, *Mémoire sur l'étude optique des mouvements vibratoires*, 2.

31 "Tous les musiciens du monde feront chanter une mesme pièce de Musique selon l'intention du Compositeur." In order to create such a standard, Mersenne proposed that the composer "give a certain and universal sign of the tone . . . if he puts in front of one of the notes of the bass, or of the other parts, the number of oscillations which make the sound." "Le compositeur donnera un signe certain et universel du ton . . . s'il met vis à vis des notes de la Basse, ou des autres parties, le

nombre des battements de l'air qui fait le son." Mersenne, *Harmonie universelle. Livre Troisième des instruments à chordes*, Proposition 23, 147.

32 "Acoustique: Sur la détermination d'un son fixe," *Histoire de l'Académie royale des sciences, année 1700* (Paris: Imprimerie royale, 1700), 134–143; and "Principes d'acoustique et de musique ou Système général des intervalles des sons, et son application à tous les systèmes et tous les instruments de musique," *Mémoires de l'Académie Royale des Sciences de l'Institut de France*, 1701, sect. XII.

33 Sauveur, "Sur les cordes sonores et sur une nouvelle détermination du son fixe," in *Histoire de l'Académie royale des sciences . . . année 1713* (Paris: Imprimerie royale, 1713), 68–75.

34 Jackson, *Harmonious Triads*, 200–205.

35 On Scheibler's tuning technique, see Jackson, *Harmonious Triads*, 151–181. On his proposed standard, see Jackson, *Harmonious Triads*, 206–207.

36 Chladni and Weber were often quoted in French acoustical works. Though less canonical, Scheibler's work also received a great deal of attention. In 1849, Alexandre Vincent published an annotated translation of Scheibler's treatise: *Mémoire sur la théorie des battements, application à l'accord de l'orgue et des autres instruments* (Paris: Bachelier, 1849); and in 1855, Augustin Lecomte, a correspondent member of the Lille Academy of Sciences, attempted to popularize his theories by publishing an easily accessible "Mémoire explicatif de l'invention de Scheibler: Pour introduire une exactitude inconnue avant lui, dans l'accord des instruments de musique," *Mémoires de l'Académie de la Société impériale des sciences de l'agriculture et des arts, de Lille* (Lille: Danel, 1856), 47–135. Scheibler's work also had an impact on the field of instrument building: Cavaillé-Coll, who met the manufacturer in Paris in 1836, reported that he became convinced of the precision of his method when he saw him present it. "De la détermination du ton normal ou du diapason pour l'accord des instruments de musique," *L'Ami de la religion et du roi* (5 Feb. 1859): 307.

37 Gaspard de Prony, *Instruction élémentaire sur les moyens de calculer les intervalles musicaux . . .* (Paris: Firmin Didot, 1832), 93–94.

38 Adrien de La Fage, *De l'unité tonique et de la fixation d'un diapason universel* (Paris: Dentu, 1859), 79.

39 *Supplément au Catalogue de Marloye* (Paris: Ducessois, 1840), n.n. In 1859, the organ builder Aristide Cavaillé-Coll wrote that Chladni's pitch "has been adopted by physicists in their books; but musicians, who at this time had gone past the limit of the proposed fixed pitch, had no consideration for it and today the *A* of artists finds itself a good quarter of a tone higher than the scholars." See "De la détermination du ton normal," 305.

40 Paolo Brenni, "19th Century Scientific Instrument Makers III. Secrétan et Lerebours," *Bulletin of the Scientific Instrument Society* 39 (1993): 3–6.

41 "Commode pour le calcul." *Catalogue et prix des instruments d'optique, de physique et de chimie, de mathématiques, d'astronomie et de marine, qui se trouvent et s'exécutent dans les magasins et ateliers de Lerebours et Secretan* (Paris: Ateliers de l'Est, 1853), 65. The adjective "normal" was commonly used to qualify scientific instruments, as attested, for instance, by its use in this same catalogue in reference to a camera objective (48) and a daguerreotype (50). In 1827, Savart

published an article entitled "Sur les vibrations normales," *Annales de Chimie et de Physique* 36 (1827): 187–208.

42 In 1856, especially, Lissajous stressed the potential his method promised for the study of the optical phenomenon of stereoscopy, which involved optical illusions by which flat images were perceived as three-dimensional. Lissajous, "Note sur un cas particulier de stéréoscopie fourni par l'étude optique des mouvements vibratoires. Tracé graphique des courbes auxquelles cette étude conduit," *Comptes rendus hebdomadaires des séances de l'Académie des sciences* 43 (17 Dec. 1856): 973–976.

43 The success of Lissajous's method in European and American classrooms, as well as other public demonstrations throughout the second half of the nineteenth century, owed much to their pedagogical appeal to a new generation of science popularizers who were eager to mobilize the sensory appeal of experimentally produced sounds to engage with diverse audiences. John Tyndall was among the main popularizers of Lissajous's technique in the second half of the nineteenth century; and the method developed by the scientist served as the basis for the commercialization of several technologies enabling the replication of the experiment. See Turner, "Demonstrating Harmony."

44 Lissajous, "Note sur l'élévation progressive du diapason des orchestres," 296.

45 Haynes, *A History of Performing Pitch.*

46 "Il fut impossible, malgré l'allongement de tous les tubes sonores de l'orchestre, de mettre la masse instrumentale d'accord avec le nouvel orgue achevé depuis trois ans." Berlioz, "Le Diapason," 3.

47 For example, Edmond Viel criticized the fact that instrumentalists and singers "were grouped around the conductor within a small space," thus rendering the perception of individual parts difficult: "instrumentistes et chanteurs, étaient groupés autour du chef, dans un espace assez restreint." *Le Ménestrel* (6 May 1855): 1. Joseph d'Ortigue similarly deplored the impression of "charivari" that resulted from the church's reverberant acoustics. "Critique musicale. L'année 1855 au point de vue de la musique religieuse. Le Te Deum de M. H. Berlioz. Une messe solennelle de M. Ch. Gounod," *Revue de musique ancienne et moderne* 1, no. 2 (Feb. 1856): 65. As often, despite these claims, Berlioz himself seemed satisfied with the performance. In a letter to Franz Liszt, he boasted how "the *Te Deum* was performed with the most magnificent precision . . . we were 950 performers. And not a single fault! I can't believe it." "le *Te Deum* a été exécuté aujourd'hui avec la plus magnifique précision . . . nous étions 950 exécutants. Et pas une faute! Je n'en reviens pas." Berlioz to Liszt, Paris, 30 April 1855. Hector Berlioz, *Correspondance Générale* V: March 1855–August 1859 [nos. 1905–2395], ed. Hugh J. Macdonald and François Lesure (1989) (Paris: Flammarion, 1989) (no. 1959).

48 Alexander Rehding, *Music and Monumentality: Commemoration and Wonderment in Nineteenth-Century Germany* (Oxford: Oxford University Press, 2009).

49 "S'il y avait, ainsi qu'on le répète souvent à Paris, une grande dissemblance entre les diapasons de l'Opéra, de l'Opéra-Comique, du Théâtre-Italien et des musiques militaires, comment eussent été possibles les orchestres de sept à huit cents musiciens qu'il m'est arrivé si souvent de diriger dans les vastes locaux des Champs-Élysées, après les expositions de 1844 et de 1855 et dans l'église de Saint-Eustache, puisque les éléments de ces congrès musicaux se composaient nécessairement de presque tous les instrumentistes disséminés dans les nombreux corps de musique de Paris?" Berlioz, "Le Diapason," 2. The composer's impression was confirmed by Cavaillé-Coll ("De la détermination du ton normal").

50 "Réunir quelquefois ces orchestres et d'en former, au moyen de certaines précautions, une grande masse instrumentale dont l'accord est satisfaisant." Berlioz, "Le Diapason," 2.

51 "Depuis un siècle environ, le diapason se serait élevé d'un ton, soit d'un demi-ton par demi-siècle, et qu'en continuant cette progression ascendante on arriverait ainsi à parcourir tous les degrés de l'échelle chromatique." Cavaillé-Coll, "De la détermination du ton normal," 302–303. This fundamental thesis of historical escalation has been dismissed in more recent work in the field of acoustics and performance practice studies. See especially Leipp and Castellengo, "Du diapason et de sa relativité"; and Bruce Haynes, *A History of Performing Pitch*. The lack of scientific and historical evidence for the claim, however, does not undermine its remarkable impact on the history of pitch negotiations.

52 "Du changement de diapason que l'on dit projeté à l'Opéra," *Revue et Gazette musicale de Paris* (23 Jan. 1840): 55. Fétis warned against "the prompt deterioration of the Opera's singers," asserting that "the consumption of this theater's personnel has arrived at this point where one will now have to renew it every five or six years," which, he contended, was "materially impossible."

53 Dolan, *The Orchestral Revolution*.

54 See, for example, the abundant use of the adjective "brilliant" in the title of musical works in French composers' catalogues. On the relation between brilliancy and certain materials, see, for instance, Johann Georg Kastner, who detailed how "the cornet is more brilliant than the bugle and the clarinet," the former being made of brass, but lower-pitched than the cornet, the second being a wood instrument. *Traité général d'instrumentation* (Paris: Philipp, 1837). Similarly, Berlioz considered that there was "nothing more brilliant" than the last notes of the saxhorn's upper octave. Hector Berlioz, *Grand Traité d'instrumentation et d'orchestration modernes* (Paris: Henry Lemoine et Cie, 1843). The Belgian composer and theorist François-Auguste Gevaert likewise emphasized how the small E-flat clarinet was "used especially in its upper register, whose sonority has a lot of brightness and shine." *Traité général d'instrumentation: Exposé méthodique des principes de cet art dans leur application à l'orchestre, à la musique d'harmonie et de fanfares, etc.* (Gand and Liège: Gevaert et Fils, 1863), 185.

55 "Le clairon a le brillant d'un soprano élevé." Gevaert, *Nouveau traité d'instrumentation* (Paris and Brussels: Lemoine et Fils: 1885), 165; and "Sur la chanterelle, le timbre du violon est brillant et éclatant." Gevaert, *Traité général d'instrumentation*, 28.

56 "Sa sonorité est plus terne, son timbre est grave et sérieux." Gevaert, *Traité général d'instrumentation*, 31–32; "il n'a pas assez . . . de brillant pour se produire avec succès dans les grands solos de *concert*." Gevaert, *Nouveau traité d'instrumentation*, 50.

57 Dolan, *The Orchestral Revolution*, chapter 6.

58 Fétis, "Du changement de diapason," 55.

59 "Si le chanteur persiste à se tenir en dehors de son étendue naturelle . . . ; il ne peut plus nuancer, enfler, diminuer le son. Une fois sur ce penchant, il ne mène plus sa voix, c'est sa voix qui le mène, et alors arrive indubitablement la monotonie, le pire des défauts, car elle est mère de l'ennui." Gustave Bénédit, *Étude artistique sur le diapason normal suivie de quelques anecdotes sur les ténors en province* (Marseille: Barlatier-Feissat et Demonchy, 1860), 27.

60 Étienne Jardin, "Le Conservatoire et la ville: Les écoles de musique de Besançon, Caen, Rennes, Roubaix, et Saint-Étienne" (PhD diss., École des hautes études en sciences sociales, 2006).

61 Archives nationales (hereafter AN), AJ/13/114.

62 The commission comprised three employees of the Paris Opera (Dubois, stage manager; Bonnemer, cashier; and Kreutzer, orchestra director), as well as three members of the Conservatory (Cherubini, director; Berton and Boieldieu, both professors), the superintendent of the Royal Chapel (Lesueur), the director of the Royal Chapel's Music (Päer), and the orchestra director of the Théâtre-Italien (Grasset) (Académie royale de musique, handwritten report by the commission of 21 June 1824. AN, AJ/13/114).

63 Académie royale de musique, handwritten report by the commission of 21 June 1824.

64 "Les membres de la commission . . . ont pensé que cette amélioration pourrait dans les premiers jours, surprendre le public en le désaccoutumant à des effets trop éclatants auxquels son oreille est habituée depuis plusieurs années"; "le diapason de l'Académie Royale de Musique ainsi rétabli, devienne en même temps celui des Théâtres lyriques royaux, de l'École Royale, et de la Chapelle." Académie royale de musique, handwritten report by the commission of 21 June 1824.

65 "Il y a longtemps que les amateurs de musique se plaignent de ce que les orchestres écrasent le chant . . . les plus belles voix s'épuiser rapidement et dépasser, pour être entendues au milieu des instruments, les limites au-delà desquelles le chant ne peut plus avoir ni grâce ni justesse d'expression ni sentiment." "Institut impérial de France—Classe des beaux-arts. Séance publique du samedi 2 octobre 1813. Notice des travaux de la classe des beaux-arts de l'Institut impérial de France pour l'année 1813 . . ." Reproduced in Jean-Michel Leniaud, *Procès-verbaux de l'Académie des beaux-arts* (Paris: École des Chartes, 12 vols., 2001–2018 [vol. 1, 2001]), 446.

66 "De jeunes élèves dont la nature n'a point encore complété les moyens, et qu'on obligeait constamment à des efforts qui usaient jusqu'au germe de leur talent." "Institut impérial de France—Classe des beaux-arts. Séance publique du samedi 2 octobre 1813," 446.

67 "Institut impérial de France—Classe des beaux-arts. Séance publique du samedi 2 octobre 1813," 446.

68 "Comparé les diapasons de tous les orchestres de Paris, et choisi un terme moyen entre ceux de l'Académie impériale de musique, de l'*Opera buffa*, de la chapelle de S. M. l'Empereur, enfin de celui de ses propres exercices, qui était aussi trop élevé." "Institut impérial de France—Classe des beaux-arts. Séance publique du samedi 2 octobre 1813," 448.

69 Académie royale de musique, handwritten report by the commission dated 21 June 1824. AN, AJ/13/114.

70 *De l'unité tonique*, 65–66. Rossini's first opera to be performed at the Paris Opera was *Le Siège de Corinthe* (premiered on 9 Oct. 1926).

71 William Weber, *The Rise of Musical*; and "La musique ancienne."

72 The physicist's lecture received a great echo in the press, from music journals to generalist titles and periodicals on instrument building.

73 The activities of this society remain difficult to document because it has left virtually no archive. Malou Haine, *Les Facteurs d'instruments de musique à Paris au XIXe siècle: des artisans face à l'industrialisation* (Brussels: Éditions de l'Université, 1985), 293.

74 Adolphe Giacomelli, "Société des fabricants de pianos. Séance du 9 juin. Unité du diapason," *La France musicale* 20, no. 24 (15 June 1856): 1.

75 Giacomelli, "Société des fabricants de pianos," 1.

76 Giacomelli, "Société des fabricants de pianos," 2.

77 Giacomelli (1825–1893) was an agent and a critic. For a very short bio, see Malou Haine, *400 lettres de musiciens au Musée royal de Mariemont* (Liège: Mardaga, 1995), 316.

78 *Traité complet et rationnel des principes élémentaires de la musique ou introduction à toutes les méthodes vocales instrumentales et à tous les traités d'harmonie* (Paris: the author, 1850). For a short biography of Bodin (1795–1862), see François-Joseph Fétis, *Biographie universelle des musiciens et bibliographie générale de la musique*, vol. 1 (2nd ed.) (Paris: Firmin Didot, 1868), 466.

79 "rien de commun avec notre échelle musicale. Il ne faut pas chercher à assimiler des choses qui n'ont pas la moindre analogie." Bodin, "Correspondance," *Le Luth français* (5 July 1856): 5.

80 "Société des fabricants de pianos. Procès-verbal de la séance du 9 juin 1856," *Le Luth français* (20 June 1856): 5.

81 The reports by *La France musicale* and *Le Luth français* diverge on the composition of the commission. In the text, I only give the list of names cited by both journals.

82 Haine, *Les Facteurs d'instruments de musique*, 293. The author does not indicate when piano makers were joined by other professions.

83 Napoleon III liberalized French theaters in 1864 to promote the nation's cultural life and was especially interested in conscripting music into state ceremonies. Jean-Claude Yon, ed., *Les Spectacles sous le Second Empire* (Paris: Armand Colin, 2010).

84 Tkaczyk, "The Making of Acoustics around 1800."

85 "Variétés. Le pouvoir et la science," *Cosmos: Revue encyclopédique hebdomadaire des progrès des sciences et de leurs applications aux arts et à l'industrie* 10, no. 1 (1857): 499–500.

86 Bruno Belhoste, "Jean-Baptiste Dumas et la promotion des sciences appliquées au XIXe siècle," in *Encourager l'innovation en France et en Europe*, 53–64.

87 "We believe we have established, Mister the Minister, that the elevation of the pitch is due to the efforts of the industry." "Nous croyons, Monsieur le Ministre, que l'élévation du diapason est due aux efforts de l'industrie." *Rapport et Arrêtés*, 10.

88 "La musique est . . . une sorte de langue universelle. Toutes les nationalités disparaissent devant l'écriture musicale, puisqu'une notation unique suffit à tous les peuples, puisque des signes, partout les mêmes, représentent les sons . . . N'est-il pas désirable qu'un diapason uniforme et désormais invariable vienne ajouter un lien suprême à cette communauté intelligente, et qu'un la, toujours le même, résonnant sur toute la surface du globe avec les mêmes vibrations, facilite les relations musicales et les rende plus harmonieuses encore?" *Rapport et Arrêtés*, 12.

89 "Il fallait commencer par nous renseigner au dehors et autour de nous," "prendre connaissance de l'état général du diapason, faire, en un mot, une sorte d'enquête." *Rapport et Arrêtés*, 11.

90 "Partout où il y a un opéra, un grand établissement musical, dans les villes où l'art est cultivé avec amour, avec succès, pratiqué avec éclat, et qu'on peut nommer les capitales de la musique." *Rapport et Arrêtés*, 11.

91 The table does not provide a value for this fork, but it is presented between two forks producing these two number of vibrations (the table is organized by decreasing number of vibrations).

92 "La France compte à ses deux extrémités un des deux diapasons les plus élevés, celui de Lille, un des diapasons les plus bas, celui de l'école de Toulouse. On peut suivre sur la carte la route que suit en France le diapason; il s'élève et s'abaisse avec la latitude. De Paris à Lille, il monte; il descend de Paris à Toulouse. Nous voyons le Nord soumis évidemment au contact, à la prédominance de l'art instrumental, tandis que le Midi reste fidèle aux convenances et aux bonnes traditions de l'école vocale." *Rapport et Arrêtés*, 20.

93 "Avec toute la précision de la science." *Rapport et Arrêtés*, 22.

94 "Le demi-ton rencontra des adversaires." *Rapport et Arrêtés*, 22.

95 "Le compositeur a dans sa tête, dans son imagination, on peut dire dans son cœur, le type naturel des voix. La phrase qu'il écrit lui est dictée par un chanteur que lui seul entend, et ce chanteur chante toujours bien. Sa voix, souple, pure, intelligente et juste est fixée d'après un diapason modéré et vrai, qui habite l'oreille du compositeur." *Rapport et Arrêtés*, 6.

96 "[Apporterait] une modération sensible aux études et aux travaux des chanteurs, sans jeter une trop grande perturbation dans les habitudes, il s'insinuerait, pour ainsi dire, *incognito*, en présence du public; il rendrait plus facile l'exécution des anciens chefs-d'œuvre, il nous ramènerait au diapason employé il y a environ trente ans, époque de la production d'ouvrages restés pour la plupart au répertoire, lesquels se retrouveraient dans leurs conditions premières de composition et de représentation." *Rapport et Arrêtés*, 24–25.

97 William Weber, *The Great Transformation of Musical Taste: Concert Programming from Haydn to Brahms* (New York: Cambridge University Press, 2008).

98 Jane Fulcher, *The Nation's Image: French Grand Opera as Politics and Politicized Art* (Cambridge: Cambridge University Press, 2002); see also Mark Everist, "The Music of Power: Parisian Opera and the Politics of Genre, 1806–1864," *Journal of the American Musicological Society* 67 (2014): 687–736.

99 "La musique religieuse, la musique dramatique ont subi le mouvement sans pouvoir s'en défendre ou sans chercher à s'y dérober." *Rapport et Arrêtés*, 10.

100 "tous les établissements musicaux . . . autorisés par l'État." *Rapport et Arrêtés*, 33.

101 Lissajous, "Note sur l'élévation progressive du diapason des orchestres"; Berlioz, "Le Diapason," 2. According to the commission, the composer's opinion was shared by Daussoigne-Méhul, Georges Hainl, and Auguste Morel. *Rapport et Arrêtés*, 12.

102 "Concilier les exigences de la science physique et des besoins de l'art musical"; "consacrer la tonalité moderne et de faire cesser la différence notable qui existe

entre le ton des physiciens et celui des musiciens. Ce rapprochement de l'art et de la science aurait, suivant nous, la plus sûre garantie pour l'adoption de la mesure et pour la conservation du diapason du dix-neuvième siècle." Cavaillé-Coll, "De la détermination du ton," 313.

103 "Un diapason unique ne saurait suffire à tous les usages musicaux." La Fage, *De l'unité tonique*, 117.

104 Bruce J. Hunt, "The Ohm Is Where the Art Is." On this question, see also Crosbie Smith and M. Norton Wise, *Energy and Empire: A Biographical Study of Lord Kelvin* (Cambridge: Cambridge University Press, 1989), 684–698; Simon Schaffer, "Late Victorian Metrology and Its Instrumentation"; by the same, "Physics Laboratories and the Victorian Country House"; and Wise, *The Values of Precision*.

105 Alder, "A Revolution to Measure."

106 "Quand on aura adopté un diapason . . . , on n'aura rien fait si l'on ne prend pas des mesures pour assurer la conservation de ce diapason et surtout son identité dans tous les orchestres de France." "Correspondance [Letter from Charles Delezenne to Adolphe Giacomelli]," *Le Luth français* (5 July 1856): 5.

107 "qu'un diapason type, exécutant 870 vibrations par seconde à la température de 15 degrés centigrades, fût construit sous la direction d'hommes compétents, désignés par Votre Excellence." *Rapport et Arrêtés*, 25–26.

108 Alder, "A Revolution to Measure."

109 "On n'aura qu'à placer dans le foyer de tous les orchestres de concert et de théâtre un tuyau d'orgue donnant exactement le son désigné." Berlioz, "Le Diapason," 2.

110 Cavaillé-Coll, "De la détermination du ton normal." The problem of organ pipes' tonal instability had been stressed by generations of acousticians since Diderot and beyond.

111 La Fage, *De l'unité tonique*, 97–98. La Fage presents the results of his research on the ancient Chinese concept of pitch in the same volume, 5–12, as well as in his *Histoire de la musique*, vol. 1, *Antiquité* (Paris: Au comptoir des imprimeurs unis, 1844). In these two publications, La Fage builds on the work of the French Jesuit missionary Jean Joseph Marie Amiot (1718–1793), who lived in China and wrote *De la musique moderne des Chinois*. Bibliothèque nationale de France, département Musique, RES VMB MS-14.

112 David Pantalony, *Altered Sensations: Rudolph Koenig's Acoustical Workshop in Nineteenth-Century Paris* (Dordrecht: Springer, 2009).

113 In addition, the architect of the Conservatory installed a bell tuned to the new standard and bearing the inscription "I give the *A* of the normal pitch instituted in 1858. I was placed on the clock of the music conservatory in 1860," to "accustom the ears" of the institution's students. Michèle Maurin, Émile Leipp, and M. Genet-Varcin, "La Cloche-diapason du Conservatoire," *Bulletin du Groupe d'Acoustique Musicale*, 88 (Dec. 1976). See also the correspondence between Jauniac, architect of the Conservatory and the Ministry of Fine Arts, held in AN, F/21/768.

114 Tristan Labouret, "L'adoption du diapason normal en France dans les années 1860: Les formations instrumentales à l'épreuve de la standardisation," *Musique-Images-Instruments* 17 (2018), *Notes et documents*, 309.

115 Letter from Théodore Dubois, head of the Paris Conservatory, to Alfred Nicolas Rambaud, minister of public instruction, fine arts, and cults (Ministre de l'Instruction publique, des Beaux-arts et des Cultes), 16 March 1898. AN, AJ/37/81.

116 Letter from Gabriel Fauré to Theodore Steeg, minister of public instruction and fine arts, Paris, 25 June 1917. AN, AJ/37/81.

117 Often inherited from the past, string instruments were unlikely to be replaced, as they sometimes were "of considerable value." Emmanuel Hervé, "Le diapason de l'Opéra de Paris," *Musique. Images. Instruments* 12 (2010): 208. The same instruments had actually been adjusted to the high pitch of the Opéra over the course of the 1820s, following a trend already attested in the first years of the nineteenth century by extending their neck and stretching their strings. Because this added pressure on the sounding board, bass bars had been replaced. After the 1859 decree, string instrument makers simply recommended changing the strings of violins, violas, cellos, and contrabasses. Hervé, "Le diapason de l'Opéra de Paris," 208.

118 Hervé, "Le diapason de l'Opéra de Paris," 208–209.

119 Letter from eighteen musicians of the Paris Opera orchestra to Girard, Paris, 21 Oct. 1859. AN, AJ/13/450.

120 Labouret, "L'adoption du diapason normal en France dans les années 1860," 305.

121 Labouret, "L'adoption du diapason normal en France dans les années 1860," 310.

122 Letter from the prefect of the north to the minister of state, 28 March 1862 AN, F/21/1238.

123 Labouret, "L'adoption du diapason normal en France dans les années 1860," 312–313.

124 Labouret, "L'adoption du diapason normal en France dans les années 1860," 313–315.

125 Jean-Yves Rauline, "Les Sociétés musicales en Haute-Normandie: Contribution à une histoire sociale de la musique" (PhD diss., Université Paris-Sorbonne, 2000), 316–321.

126 Labouret, "L'adoption du diapason normal en France dans les années 1860," 316–324.

CHAPTER TWO

"Die neue Orchesterstimmung, obwol sie nicht den geistigen, sondern nur den technischen Theil der Kunst angeht, bleibt immerhin ein Capitel der musikalischen Culturgeschichte. Nach rechts und links, von Paris bis Petersburg hat sie ihre Fäden ausgebreitet und wird bald als musikalisches Eisenbahnnetz alle Culturstädte verbinden." Eduard Hanslick, "Musikalische Briefe," *Neue Freie Presse* (10 Sept. 1864), in Eduard Hanslick, *Sämtliche Schriften. Historisch-kritische Ausgabe, Band I/7, Aufsätze und Rezensionen 1864–1865*, ed. Dietmar Strauß (Vienna, Cologne, and Weimar: Böhlau, 2011), 138.

1 Mersenne, *Harmonie universelle. Livre Troisième des instruments à chordes*, Proposition 18, 147.

2 Bernard Dompnier, "Les maîtrises capitulaires et l'apprentissage du métier de musicien dans la France d'Ancien Régime," in *La Tradition du savoir*, ed. Guy

Bedouelle, Christian Belin, and Simone de Reyff (Fribourg: Academic Press, 2011), 131–151.

3 Along with sight reading, a good quality of tone, and the ability to perform music expressively. The examination reports of the institution's directors reflect these new expectations: see for example *École de Musique et de Déclamation. Examens des Classes de l'École, par les membres du Comité. Registres de Mr. Cherubini, Directeur*, 4 vols., 1816–1834. AN, AJ/37/207/4 and 208/1–3. I thank Étienne Jardin for sharing this source with me. On the case of singers' evaluation, see Pierre Girod, "Les Mutations du ténor romantique. Contribution à une histoire du chant français à l'époque de Gilbert Duprez (1837–1871)" (PhD diss., Université européenne de Bretagne, 2015), 280–282.

4 See William E. Echard, *Napoleon III and the Concert of Europe* (Baton Rouge: Louisiana State University Press, 1983). On Anglo-French relations, see David Brown, "Palmerston and Anglo-French Relations, 1846–1865," in Glyn Stone and T. G. Otte, eds., *Anglo-French Relations Since the Late Eighteenth Century* (London and New York: Routledge, 2008), 41–58.

5 Similar arguments have been made about France and its role in the field of astronomy. See Lorraine Daston, "The Immortal Archive: Nineteenth-Century Science Imagines the Future," in Lorraine Daston, ed., *Sciences in the Archives: Pasts, Presents, Futures* (Chicago: University of Chicago Press, 2017), 159–182.

6 "Die Kammertonfrage kann man in gewissem Sinne eine kosmopolitische nennen, weil bei der Erörterung von der Möglichkeit einer Reducirung der modernen Orchesterstimmung die ganze musikalische Welt interessirt ist." Dr. J, "Die Reform der Orchesterstimmung," *Süddeutsche Musik-Zeitung* 7, no. 42 (18 Oct. 1858): 166.

7 (Anon.), "An Imperial Pitchfork," *The Spectator* (28 Aug 1858): 910–911.

8 "Die Herabsetzung der Orchesterstimmung," *Süddeutsche Musik-Zeitung* 9, no. 50 (10 Dec. 1860): 198.

9 "Tuning-Fork and Musical Pitch," *Chamber's Journal of Popular Literature Science and Arts* 34 (18 Aug. 1860): 99. Quoted in Edward Gillin, *Sound Authorities: Scientific and Musical Knowledge in Nineteenth-Century Britain* (Chicago: University of Chicago Press, 2021).

10 I selected these prominent examples not only because of Britain's, Austria's, and Germany's centrality in the European concert, but also because they are exceptionally well documented.

11 This analysis of Britain's reception of the French pitch draws on the following article: Edward Gillin and Fanny Gribenski, "The Politics of Musical Standardization in Nineteenth-Century France and Britain," *Past and Present* 261, no 1 (2021): 153–187.

12 Haynes, *A History of Performing Pitch.*

13 David Brown, "Palmerston and Anglo-French Relations."

14 Derek Hudson and Kenneth W. Luckhurst, *The Royal Society of Arts 1754–1954* (London: Murray, 1954).

15 "Tuning-Forks and Musical Pitch," 100–101.

16 Royal Society of Arts Library, London (hereafter RSA), AD/MA/100/12/02/11. *Minutes of the Council*, 105 (11 May 1859).

17 *Minutes of the Council,* 112 (25 May 1859); and 122 (1 June 1859).

18 The lawyer Thomas Philipps (chairman), William Whewell, Augustus de Morgan, Henry Wylde, Charles Wheatstone, and Robert Willis offered scientific expertise, while musical representatives included the composers Sterndale Bennett and Henry Griesbach, the music critic Henry Lunn, and the acclaimed conductor George Smart. There were also celebrated instrument makers, such as Walter Broadwood and Henry Willis.

19 "Uniform Musical Pitch: Minutes of a Meeting of Musicians, Amateurs, and Others Interested in Music, Held at the House of the Society of Arts, when the Report of the Committee Appointed by the Council of the Society Was Received and Adopted," *Journal of the Society of Arts* 8, no. 417 (16 Nov. 1860): 3.

20 "Introduction: State and Society in Victorian Britain," in Peter Mandler, ed., *Liberty and Authority in Victorian Britain* (Oxford: Oxford University Press, 2006), 1–2; and Martin Daunton, *State and Market in Victorian Britain* (Woodbridge, UK: Boydell Press, 2008), 4–5.

21 "Uniform Musical Pitch: Minutes," 4.

22 Letter from W. Watson to Peter le Neve Foster, Lincoln, 6 Sept. 1859. RSA PR/GE/121/10/5.

23 Letter from W. Lockyer to Foster, 10 Sept. 1859; and Letter from Charles Saldman to Foster, London, 10 Sept. 1859. RSA PR/GE/121/10/5.

24 On the history of *C* 512, from Sauveur to Chladni, de Prony, and Marloye, see above, chapter 1.

25 John Herschel, "Uniform Musical Pitch," *Leeds Mercury* (2 Aug. 1859).

26 "Uniform Musical Pitch: Minutes," 4.

27 "Uniform Musical Pitch: Minutes," 4–5.

28 "Uniform Musical Pitch: Minutes," 5.

29 "Uniform Musical Pitch: Minutes," 5.

30 Henry C. Lunn, "Musical Pitch," *The Musical Times and Singing Class Circular* 13, no. 312 (1 Feb. 1869): 663–665.

31 Haynes, *A History of Performing Pitch,* 356–357.

32 Michael Nagy, "Historisches Umfeld und Entwicklung der Wiener Oboe, "in *Zur Situation der Musiker in Österreich: Referate der Musik-Symposium im Schlosshof 1989–93,* ed. Paul W. Fürst (Vienna: Institut für Wiener Klangstil, 1994), 140; Letter from the K. K. Oberst Kämmerer Amt to the Ministry of Foreign Affairs, Vienna, 10 June 1862. AT-OeStA:HHStA MdÄ AR F60 35–1. "Orchester-Stimmung. Akten, 1862–1869." Vienna, Staatsarchiv (hereafter "Orchester-Stimmung. Akten").

33 Circular letter from Vesque to the Imperial- Royal Austrian diplomatic missions in Germany, Vienna, 13 May 1862. AT-OeStA:HHStA MdÄ AR F60 35–1. "Orchester-Stimmung. Akten."

34 "Austria," in *New Grove Dictionary of Music and Musicians,* ed. Stanley Sadie, 2nd ed. (London: Macmillan, 2001).

35 Deborah R. Coen, "Climate and Circulation in Imperial Austria," *Journal of Modern History* 82, no. 4, *Science and the Making of Modern Culture* (December 2010): 865.

36 Catherine Horel, "La France et l'Empire d'Autriche, 1815–1918," *Bulletin de l'Institute Pierre Renouvin* 32, no. 2 (2010): 57–79.

37 Historian of science Deborah R. Coen has shown how the emergence of climate science in Austria was part of the country's efforts to secure unity across their territory. See "Climate and Circulation in Imperial Austria"; and *Climate in Motion: Science, Empire, and the Problem of Scale* (Chicago: University of Chicago Press, 2019).

38 On the foreign policy of the Habsburg monarchy throughout the nineteenth century, see especially: F. R. Bridge, *The Habsburg Monarchy: Among the Great Powers, 1815–1918* (Providence, RI: Berg, 1990); Adam Wandruszka and Peter Urbanitsch (eds.), *Die Habsburgermonarchie im System der internationalen Beziehungen*, in *Die Habsburgermonarchie 1848–1918*, vol. 6 (Vienna: Österreichische Akademie der Wissenschaften, 1989–1993); Helmut Rumpler, *Eine Chance für Mitteleuropa. Bürgerliche Emanzipation und Staatsverfall in der Habsburgermonarchie, Österreichische Geschichte 1804–1914* (Vienna: Ueberreuter, 1997); and Catherine Horel, "Du congrès de Vienne au traité de Saint-Germain. L'Autriche et le Concert européen, 1815–1919," *Bulletin de l'Institute Pierre Renouvin* 42, no. 2 (2015): 33–46.

39 Coen, *Climate in Motion*, 54–55.

40 In the aftermath of Napoleon I's military expansion, the Congress of Vienna established the principle of a balance of powers within Europe. The creation of a German confederation was part of these efforts to prevent the constitution of too large territorial entities placed under the authority of a sole sovereign in Europe.

41 Peter Katzenstein, *Disjoined Partners: Austria and Germany since 1815* (Berkeley: University of California Press, 2018 [reprint: 1976 for the 1st ed.]).

42 Katzenstein, *Disjoined Partners*, 84–85.

43 Katzenstein, *Disjoined Partners*, 79–80.

44 "Offenbar wollte unsere Regierung sich von dem centralisirenden Despotismus Frankreichs ferne halten, der im Verordnungswege das ganze Reich, Theater, Concerte, Schulen und Fabriken binnen Jahresfrist unter Eine Stimmgabel gebracht haben . . . die wichtigsten Musikinstitute Oesterreichs [hätten] immerhin die neue Stimmungseinheit aus freien Stücken adoptiren können. Jetzt noch ließe sich auf dem gütlichen Wege der Belehrung und Anempfehlung für die Einheit der neuen Stimmung Vieles thun." Hanslick, "Musikalische Briefe," 138.

45 Hanslick, "Musik. (Hofoperntheater.–Orchester.–Die Pariser Stimmung)," Presse, 7 Oct. 1861, in Hanslick, *Sämtliche Schriften. Band I/5*, 390.

46 Circular letter from Vesque to the Imperial-Royal Austrian diplomatic missions in Germany, Vienna, 13 May 1862. "Orchester-Stimmung. Akten."

47 An 1848er, Anton von Schmerling (1805–1893) briefly served in the government elected by Austria's new Parliament after the Revolution. Having resigned soon after, in the context of growing neo-absolutism, von Schmerling came back to power as president-minister of Austria's first liberal government in 1860, and as state minister, from 1861 to 1865.

48 Circular letter from Vesque to the Imperial-Royal Austrian diplomatic missions in Germany, Vienna, 13 May 1862. "Orchester-Stimmung. Akten."

49 Katzenstein, *Disjoined Partners*, 77.

50 "ob die Königliche Regierung geneigt ist, die französische Orchesterstimmung einzuführen." Circular letter from Meysenberg (for the foreign minister) to the Imperial-Royal Austrian diplomatic missions in Germany, Vienna, 13 May 1862. "Orchester-Stimmung. Akten."

51 Circular letter from Meysenberg.

52 "Durch exacte physikalische Operate." Circular letter from Meysenberg.

53 "die längere Erhaltung der Sänger und Sängerinnen." Circular letter from Meysenberg.

54 "die ganze musikalische Welt hat mit lebhafter Zustimmung diese, von den Tonkünstlern lang ersehnte Reform begrüsst." Circular letter from Meysenberg.

55 "Durch die in letzterer Zeit erfolgte wunderbare Vermehrung früher unbekannter CommunicationsMittel, welche die Entfernungen, Kosten und Beschwerden der Reisen von ehedem auf ein noch vor Kurzem nicht geahntes Mass herabgesetzt haben, wird nämlich auch der Verkehr zwischen den musikalischen Hauptstädten Europa's ein alljährlich lebhafterem und wird es immer häufiger stattfinden, dass Wiener Sänger in Berlin, Paris, St. Petersburg, London gastieren, und umgekehrt." Circular letter from Meysenberg.

56 Letter from Bismarck to Heinrich von Mühler; Letter from von Mühler to Bismarck, 10 April 1863. "Zur Einführung einer gleichmäßige Orchesterstimmung (1862–1886). 901–70190. Bundesarchiv (hereafter "Zur Einführung einer gleichmäßige Orchesterstimmung").

57 Letter from Platen Hallermund to von Ingelheim, Hannover, 15 Nov. 1862. "Orchester-Stimmung. Akten."

58 Letter from von Cahimst to von Lützow, Darmstadt, 8 Nov. 1862. "Orchester-Stimmung. Akten."

59 Letter from Platen Hallermund to von Ingelheim, Hannover, 15 Nov. 1862. "Orchester-Stimmung. Akten."

60 Letter from Schmidt to Gustav von Blome, Bremen, 2 June 1862. "Orchester-Stimmung. Akten."

61 Letter from Benot to von Werner, Dresden, 25 July 1862. "Orchester-Stimmung. Akten." During the month of September, however, Dresden was the site of a meeting of Kapellmeisters from across Germany who decided to adopt the French pitch, despite the local Kapellmeister Karl Näke's opposition to the reform. See Karl Näke, *Ueber Orchesterstimmung. Den deutschen Kapellmeistern bei ihrer Versammlung in Dresden, den 28. September 1862 gewidmet* (Dresden: Liepsch and Reichardt, 1862); and Jackson, *Harmonious Triads*, 215–217.

62 Letter from Raggenbach (President of the grand ducal ministry) to von Pilat, Carlsruhe, 19 July 1862. "Orchester-Stimmung. Akten."

63 Letter from von Grüner (Legationsrat in Lepzig) to Rechberg, 12 July 1862; Letter from von Grüner to Rechberg, Leipzig, 22 Sept. 1862. "Orchester-Stimmung. Akten."

64 Nils Brunnsson and Andreas Rasche, "The Dynamics of Standardization: Three Perspectives on Standards in Organization Studies," *Organization Studies* 33, no 5–6 (2012): 613–632.

65 Letter from Schmidt to von Blome, Bremen, 2 June 1862. "Orchester-Stimmung. Akten."

66 Letter from Schmidt to von Blome, Bremen, 2 June 1862; Letter from von Grim to von Grüner, Greiz, 4 July 1862. "Orchester-Stimmung. Akten."

67 Erich Tremmel, *Blasinstrumentenbau im 19. Jahrhundert in Südbayern* (Augsburg: Wißner, 1993), 213–215.

68 BayHStA, Gesandtschaft Paris 10620. Bayerisches Haupstaatsarchiv (https://dfg-viewer.de/show/?tx_dlf%5Bid%5D=https://gda.bayern.de/mets/4f4a70dd-9189-443f-a727-e175e7a9cbc9, accessed 25 March 2020).

69 "Proceedings of the Society. Musical Pitch," *Journal of the Society of Arts* 17, no. 870 (23 July 1869): 699–702.

70 Haynes, *A History of Performing Pitch*, 351–366; Exposé zu der Eingabe der Gesellschaft der Musikfreunde und Genossen in Wien an das k. k. Cultus- und Ministerium, betreffend die Herbeiführung einer einheitlichen musikalischen Normalstimmung, 4. "Zur Einführung einer gleichmäßige Orchesterstimmung."

71 "Alexey Lvov," *New Grove Dictionary*.

72 *Rapport et Arrêtés*, 14.

73 Greece adopted the diapason normal as a national standard in 1910. Stella Kourmpana, "Delving into the Athens Conservatoire Archive: Musical Education as a National Need," in Polina Tambakaki, Panos Vlagopoulos, Katerina Levidou, and Roderick Beaton (eds.), *Music, Language and Identity in Greece: Defining a National Art Music in the Nineteenth and Twentieth Centuries* (London: Routledge, 2019), 104.

74 Michael Fend and Michel Noiray (eds.), *Musical Education in Europe (1770–1914): Compositional, Institutional, and Political Challenges* (Berlin: Berlin Wissenschafstverlag, 2005).

75 Alexander J. Ellis, "On the History of Musical Pitch," *Journal of the Royal Society of Arts* 28, nos. 1424 and 1428 (1880): 293–336 and 400–403; Ellis, "The History of Musical Pitch in Europe," in Hermann von Helmholtz, *On the Sensations of Tone as a Physiological Basis for the Study of Music*, 3rd ed. (London: Longmans, Green, and Co., 1895), 493–513. Both texts are an expansion of this initial paper: "On the Measurement and Settlement of Musical Pitch," *Journal of the Society of Arts* 25, no. 1279 (1877): 664–687. As Ellis explains in his texts, his work on musical pitch was inspired by the Society of Arts' 1869 initiative to gather an international collection of tuning forks in order to investigate the question of pitch standardization on a global scale.

76 On Helmholtz's contribution to acoustics, see especially Steege, *Helmholtz and the Modern Listener*; Hui, *The Psychophysical Ear*; and Julia Kursell, *Epistemologie des Hörens: Helmholtz' physiologische Grundlegung der Musiktheorie* (Paderborn: Wilhelm Fink Verlag, 2018).

77 Julia Kursell, "Alexander Ellis's Translation of Helmholtz's *Sensations of Tone*," *Isis* 109, no. 2 (2018): 339–345.

78 On the current life of this standard, see below.

79 "Il y a lieu, non d'adopter la décision française, mais de la réviser." Meerens, *Le Diapason et la notation musicale simplifiés*, 20.

80 Meerens, *Le Diapason et la notation musicale simplifiés*, 15.

81 Meerens, *Le Diapason et la notation musicale simplifiés*, 17.

82 "Pourquoi les musiciens voudraient-ils que le mètre, par exemple, qui constitue le principe de tous les poids et mesures, vaille exceptionnellement pour eux un mètre et une fraction, comme 1m05?" Meerens, *Le Diapason et la notation musicale simplifiés*, 19.

83 Meerens, *Le Diapason et la notation musicale simplifiés*, 13.

84 "Per esigenze matematiche"; "la differenza è così piccola, quasi impercettibile all'orecchio." Letter from Verdi to the president of the Commissione Permanente musicale e drammatica del Ministero della Pubblica Istruzione, 10 Feb. 1884. Quoted in Marcello Conati, "Teatri e orchestra al tempo di Verdi," in *Giuseppe Verdi: vicende, problemi e mito di un artista e del suo tempo* (Colorno, Edizioni "Una Città costruisce una mostra," 1985), 70.

85 "Aufruf zur Einführung einer allgemeinen Normalstimmung in Deutschland," *Zeitschrift für Instrumentenbau* 4(1884):363–366. Quoted in Jackson, *Harmonious Triads*, 218.

86 "weil durch eine einseitige, nur auf Oesterreich beschränkte Regelung derselben nur noch mehr Verwirrung und Incongruenz herbeigeführt würde, die Austragung der Angelegenheit im Wege einer internationalen musikalischen Conferenz." Letter from the Imperial-Royal Austrian-Hungarian embassy in Berlin, 26 April 1885. "Zur Einführung einer gleichmäßige Orchesterstimmung."

87 Letter from the Imperial-Royal Austrian-Hungarian embassy in Berlin.

88 Kaiserlich-Königliches Ministerium für Cultus und Unterricht, *Beschlüsse und Protokolle der Internationalen Stimmton Conferenz in Wien 1885* (Vienna: Kaiserlich-Königlicher Schulbücher, 1885).

89 Draft of a letter to Freycinet, Paris, 7 Oct. 1885. "Diapason Normal." AT-OeStA/HHStA Gesandtschaftsarchiv Paris 349–2. Additionally, the participants could consider simplifying international acoustical communication by unifying the various units involved in pitch quantification. Attesting to an intensification of transnational exchanges within the acoustic field, the conference acknowledged the difficulties arising from the lack of standard denominations for musical notes, coexistence of simple and double vibrations, and use of competing units for the measure of temperature (Celsius and Fahrenheit).

90 "[Der praktische Musiker] kann sich mit dem einen *A* so gut befreunden mit dem anderen; er wünscht nur ein *A*, das er als Sänger leicht und effectvoll singt, das er als Instrumentalist klangvoll spielt"; "Die Frage des Diapasons sei keine blos musikalische, sie sei auch eine Frage der Theorie"; "Italien gar keine Einwendung dagegen hätte einen anderen Diapason als den zu 864 Schwingungen anzunehmen, vorausgesetzt, dass dieser Diapason einstimmig angenommen wurde, oder wenigstens eine genügend starke Majorität für sich habe." *Beschlüsse und Protokolle*, 13–14 and 15.

91 Letter from Giuseppe Verdi to Arrigo Boito, Sant' Agata, 8 Nov. 1885, reproduced in Marcello Conati and Mario Medici (eds.), *The Verdi-Boito Correspondence*, trans. William Weaver (Chicago: University of Chicago Press, 1994), 89–90.

92 *Beschlüsse und Protokolle*, 16.

93 Jackson, *Harmonious Triads*, 224.

94 "L'emploi du diapason normal n'étant imposé par aucune loi, il n'y a pas lieu de nommer un inspecteur chargé de veiller à l'emploi régulier du diapason normal.

La Com^{on} dont il est fait mention dans l'arr^{é} min^{el} du 16 fev. 59 (et non dans la loi du 16 fev. 59) n'a jamais fonctionné pour la vérification des inst^{ts}. Le Diap. norm. N'a été imposé qu'aux écoles nat^{les}, aux théâtres subventionnés, aux mus. mil^{res}, en un mot à tous les établissements placés sous la dépendance de l'État. En conséquence, les facteurs d'inst^{ts} l'ont adopté de leur côté et l'ont pris pour base dans leur fabrication. Les diapasons soumis au poinçonnage sont vérifiés par M. Alb. Sandoz, professeur de physique. Ce prof.^{r} reçoit pour ce service une indemnité du Cons.re. Il est imposs. D'affirmer que le diap. norm. soit exclus.^{t} employé en France." Draft and copy of a letter from Thomas to the minister of education and fine arts, Paris, 28 May 1885. AN, AJ/37/81.

95 On these experiments, see George Biddell Airy, *On Sound and Atmospheric Vibrations with Mathematical Elements of Music* (London: Macmillan and Co., 1868).

96 *Beschlüsse und Protokolle*, 29.

97 *Beschlüsse und Protokolle*, 7.

98 "La température moyenne des salles où se font d'ordinaire les séances de musique, salles de concerts ou théâtres, dépasse rarement 20 degrés surtout aujourd'hui, où par des considérations d'hygiène et de sécurité, on tend de plus en plus à se servir de l'éclairage électrique qui ne modifie pas sensiblement la température d'une salle." Draft of a letter from Thomas to the ministry of public instruction and fine arts, Paris, 24 Jan. 1887. AN, AJ/37/81.

99 *Beschlüsse und Protokolle*, 6.

100 Jackson, *Harmonious Triads*, 226–227; and Miriam Focaccia, *Pietro Blaserna and the Birth of the Institute of Physics in Rome: A Gentleman Scientist at Via Panispera* (Cham: Springer, 2019).

101 On Case Western, see below, chapter 3. On Moscow, see Pantalony, *Altered Sensations*, 141.

102 The world-leading authority in the realm of acoustical instrument making from 1860 to his death in 1901, Koenig became the official producer of standard forks in France after the retirement of Secrétan and Lerebours, who had manufactured the first prototype of the diapason normal. Both attesting and contributing to Paris's centrality in the world of scientific instrument making, Koenig exported his products throughout Europe, North America, and beyond. On this figure, see Pantalony, *Altered Sensations.*

103 Pantalony, *Altered Sensations*, 100–105.

104 A standard tuning fork *A* 870 at 15°C on a tripod with resonator, and a tuning fork 870 on a resonance box. 1897 Inventory. Fakultät für Physik, Universität Wien.

105 "Staatliche Kommission für die Beglaubigung von Stimmgabeln." A "prototype" and a tuning fork whose pitch was set two vibrations higher, at *A* 437. "Staatliche Kommission für die Beglaubigung von Stimmgabeln"; draft of a letter to the ministry of cults and education [Kultus und Unterricht], n.d. B 23–238/3. Fakultät für Physik, Universität Wien.

106 The set purchased by Blaserna included three forks with brass resonators fixed on a cast-iron tripod stand, respectively giving *C* 512, *A* 435, and *B* 921.7—calculated from to *A* 435, this last fork served as a reference for the tuning of wind instruments. Blaserna secured an extraordinary funding of 15,000 lira for these initial purchases. The instruments of the *Uffici* are held at the Museum of Physics at the

University La Sapienza, and the collection is presented in this book: Maria Grazia Ianniello, *La storia dell'Istituto di Fisica della Sapienza attraverso le sue collezioni di strumenti* (Rome: Museo di Fisica di Roma, 2003). On Lowenherz's recommendations, see Jackson, *Harmonious Triads*, 226–227.

107 Pantalony, *Altered Sensations*, 25.

108 Nazzareno Pierpaoli, "Influenza della temperatura sul numero delle vibrazioni d'un corista. Nota 1," *Rendiconti della Reale accademia dei Lincei*, vol. 4 (1888), 714–718; "Influenza della temperatura sul numero delle vibrazioni d'un corista. Nota 1," *Rendiconti della Reale accademia dei Lincei*, vol. 5 (1889), 265–268; and *Studio dei coristi prototipi affidati all'Ufficio centrale italiano per il corista uniforme. Memoria* (Rome: Reale accademia dei Lincei, 1892), 200–215.

109 In 1889, for instance, in preparation for the opening of the Vienna office, Stefan reported his correspondence with the Parisian maker to the minister of education. Draft of a letter from Stefan to the minister of cults and education [Kultus und Unterricht], 1889. B23–238/3. Fakultät für Physik, Universität Wien.

110 The physicists Stefan and Victor von Lang, both professors of physics at the University in Vienna, sat with Josef Hellmesberger, first Kapellmeister and director of the Conservatory of the Gesellschaft der Musikfreunde in Wien, Leopold August Zellner, general secretary of the Gesellschaft der Musikfreunde, as well as the vice secretary of the ministry of cults and education [Kultus und Unterricht], Wilhelm Freiherr von Weckbrecker.

111 "Für *Präcisions-Gabeln* ist diese Fehler Grenze mit 0.2 einfacher (0.1 ganzen) Schwingung," "als zwischen 869.8 u 870.2 einfachen Schwingungen festzusetzen"; "Die Fehler-Grenze, innerhalb welche bezüglich der ersteren Kategorie (Gebrauchs-Gabeln) die Beglaubigung . . . stattfinden kann, ist mit 1 einfachen Schwingung (1/2 ganzen Schw.) mehr oder weniger den Sollwerten des Normaltones . . . zwischen 869 und 871 einfachen Schwingungen zu fixieren." Aus dem Protokoll über die am 9. Februar 1891 erste Besprechung der Kommission. Handwritten note. B 23 238/3. Fakultät für Physik, Universität Wien.

112 Letter from Fritz Wagner to the Imperial-Royal University in Vienna, Graz, 31 Jan. 1891. "Korrespondenz von 20/9/1890–25/2/91. Protokoll des Stimmgabelverificationsamtes für 1891. Zahl 1–300." Felix Ehrenhaft Stimmgabelnormung. Fakultät für Physik, Universität Wien (hereafter "Protokoll des Stimmgabelverificationsamtes").

113 "Musikübungen und Aufführungen und bei Kirchlichen Gelegenheiten." Letter from Bürgermeister Gutenstein to the Physikalisches Cabinet der Universität in Wien, 30 Dec. 1892. "Protokoll des Stimmgabelverificationsamtes."

114 "wann die Stimmgabeln einzusenden sind." Letter from Houdek & Hervert to Lang, Prague, 19 Feb. 1891. "Protokoll des Stimmgabelverificationsamtes."

115 Letter from Alois Kreidl to the Physics Cabinet of the University in Vienna, Prague, 9 Nov. 1891. "Protokoll des Stimmgabelverificationsamtes."

116 "Protokoll über die am 2 Mai 1912, 12h a.m.im I. physikalische Institute (früher physikalisches Kabinett) der k. k. Universität Wien abgehaltenen zweiten Besprechung der Stimmgabel-Verifikation-Kommission." B 23–258. Fakultät für Physik, Universität Wien.

117 *Bericht des Obmannes der Abteilung für Musikinstrumente des niederösterreichischen Gewerbereines Herrn Teofil Kotykiewicz, Harmoniumfabrikant, als Mitglied der Kommission zur Kontrolle der Normalstimmung.* "Protokoll des Stimmgabelverificationsamtes."

118 "Der Grund dieser Neigung liegt in der Eigentümlichkeit der Verhältnisse der Orchesterinstrumente. Im Gegensatz zu der Klavierinstrumenten ist das Orchester kein 'temperierter' Klangkörper, sondern, seiner Anlage nach, ein 'rein' ausgestimmter. In der Praxis bedeutet das, dass jedes Instrument den jeweiligen Akkordton, den es zu spielen hat, so nimmt, dass er sich in absoluter Tonhöhe in die Harmonie einfügt. Wie dies zu geschehen hat, liegt ganz im Empfinden des Instrumentalisten und ist weder im physikalisch-wissenschaftlichen, noch mit akustischen Messapparaten zu erreichen, sondern nur durch ein lebhaftes musikalisches Gefühl, das bei jedem Musiker die unbedingte Voraussetzung sein muss." Alexander Wunderer, "On the Question of Normal Pitch," *Die Oboe*, no. 6 (Nov. 1929).

119 "Um diese absolute Reinheit herstellen zu können, muss jedes Orchesterinstrument in der Tonhöhe gewissermassen 'labil' sein, das heißt jeder Ton muß die Fähigkeit haben, nach der Tiefe und nach der Höhe biegsam zu sein." Wunderer, "On the Question of Normal Pitch."

120 In Melbourne and Sydney, conversations about pitch standards started following the reprint of a British article from 1858. "Pitch Reform in France and England," *Sydney Morning Herald*, 24 June 1859, 8; "Pitch Reform in France and England," *Courier*, 20 May 1859. Quoted in Simon Purtell, *Tuning the Antipodes: Battles for Performing Pitch in Melbourne* (Melbourne: Lyrebird Press, 2016).

121 Lothar von Falkenhausen, *Suspended Music: Chime Bells in the Culture of Bronze-Age China* (Berkeley: University of California Press, 1993).

122 Western scholars knew of feudal China's traditions and instruments through the work of Jesuit Joseph Marie Amiot (on the connection between this figure and music theorist Adrien de La Fage, see above, chapter 1).

123 In Melbourne, Britain's high standard prevailed in army bands until the 1970s, long after the adoption of *A* 435 for operatic and instrumental practice in the city and the publication of the norm ISO 16 fixing the tuning frequency at *A* 440 across the world. Purtell, *Tuning the Antipodes*, 57–80.

124 Gillin, *Sound Authorities.*

125 Gillin, *Sound Authorities.*

126 "A National Musical Pitch. Reasons for adopting the 'Regulation' military band pitch, already established at the school of music, Kneller Hall, viz., C542, or A455 vibrations, 10 May, 1886," RSA PR/GE/121/10/6. Quoted in Gillin, *Sound Authorities.*

127 Alfred J. Hipkins, "The Standard of Musical Pitch," *Journal of the Society of Arts* (28 Feb. 1896): 34.

128 See below, chapter 3.

129 David J. Blaikley, "The Musical Pitch," *St James's Gazette*, 11 July 1899. The article was also published in *The Standard* (15 July) and the *Times* (16 July), as well as in this repository of most articles related to the 1899 debates initiated by piano

makers, *Musical Pitch. Letters, Articles and Comments in the Press of the Proposal to Adopt the Low Pitch Throughout the Pianoforte Trade. Reprinted for the information of all interested in the question* (London: Waterlow and Sons, 1899), 21.

130 David Brian Hammond, "British Army Music in the Interwar Years: Culture, Performance, and Influence" (PhD diss., The Open University, 2018), 202–223.

131 John Darwin, *The Empire Project: The Rise and Fall of the British World System. 1830–1970* (Cambridge: Cambridge University Press, 2009).

132 Gustave Flaubert, *Carnets de voyage, no 4: notes prises au court d'un voyage en Orient*, 24 Dec. 1849 entry. Quoted by Douglas Bush and Richard Kassel (eds.), *The Organ: An Encyclopedia* (New York and London: Routledge, 2006), 20.

133 *The Organ: An Encyclopedia*, 20–21, 35–36, 101–104.

134 Carmel Raz, "How the Sheng Became a Harp," *Sound Studies* 6, no. 2 (2020): 239–256, doi:10.1080/20551940.2020.1794648.

135 French Catholic missionaries brought harmoniums to China before 1860, quickly followed by British Protestant ones. See Ludovic des Garets, *Lettres de Chine, 1859–1861. Campagne de Chine et de Cochinchine*, ed. Geneviève Deschamps, Odile Bach, and Thierry des Garets (Gleizé: Éditions du Poutan, 2014), 118; Rev. W. E. Soothill, "Chinese Music and Its Relation to Our Native Services," *The Chinese Recorder* 21 (1890): 221–228, 336–337; and Walter Moule, "In Memoriam. Thomas Herbert Harvey, M. A.," *The Chinese Recorder* 21 (1890): 490–492. The instrument's presence is also attested in Japan after 1880: Rev. W. P. Buncombe, "An Eight Day's Mission at Tokushima," *The Church Missionary Intelligencer* (London, 1889): 430; and in Tunisia from 1871 onward: Yassine Guettat, "Le *mālūf* tunisien: origines et mutations," *RTM Revue des Traditions Musicales* 12, "Musicologie francophone du Maghreb. Mélanges offerts à Mahmoud Guettat," 77–97. I warmly thank François Picard for pointing these sources and references to my attention.

136 Claudius Torp, *Klavierwelten: Aufstieg und Verwandlung einer europäischen Kultur, 1830–1940* (Frankfurt am Main: Campus, 2022).

137 Trevor Herbert and Helen Barlow, *Music and the British Military in the Long Nineteenth Century* (Oxford: Oxford University Press, 2013), 240–268.

138 See for instance: "It may be of use to our American piano manufacturers to know that if they wish to do any business successfully in the tropics, their cases must be made of solid mahogany . . . varnish work is not thought so well of in the Far East as is the French polish or the Japanese lacquer work." "George P. Bent Tells of Pianos in the Tropics," *The Music Trade Review* 73, no. 2 (1921): 17; and "American Pianos Can Be Sold to Latin America," *The Music Trade Review* 89, no. 8 (1930): 4–5.

139 Superintendent of Documents, *Commerce Reports* 2, no. 21 (26 May 1924): 506.

140 James D. Blaikley, "An Essay on Musical Pitch," in Charles Russel Day, *A Descriptive Catalogue of the Instruments Recently Exhibited at the Royal Military Exhibition, London, 1890* (London: Eyre and Spottiswoode, 1890), 248.

141 Blaikley, "An Essay on Musical Pitch," 248.

142 In 1918, the US Army and Navy introduced pitch and temperature specifications according to which all instruments "must give A-435 at 59°F, and A-440 at 71° or 72°." Commenting on this strategy, the piano tuner D. A. MacDonald, from the Chicago-based and flourishing firm Lyon & Hailey, mocked this approach, referring to Blaikley to remind his audience that "If one instrument could be made to

do this—and I do not consider this possible—the others could not comply with this requirement." "The Finished Piano and the Tuner," in Acoustic Department. American Steel & Wire Company, *Piano Tone Building. Proceedings of the Piano Technicians Conference. Chicago 1916, 1917, 1918. New York 1919* (New York: American Steel & Wire Company, 1919 [Chicago Technician Conference of 3 April 1918]), 180.

143 Kofi Agawu, "Tonality as a Colonizing Force in Africa," in Radano and Olaniyan, *Audible Empire*.

144 Radano and Olaniyan, *Audible Empire*, "Introduction," 5.

145 Matt Rahaim, "That Ban(e) of Indian Music: Hearing Politics in the Harmonium," *Journal of Asian Studies* 70, no 3 (2011): 657–682; and Amanda Weidman, *Singing the Classical, Voicing the Modern: The Postcolonial Politics of Music in South India* (Durham, NC: Duke University Press, 2006).

146 Quoted in Rahaim, "That Ban(e) of Indian Music," 669.

147 Quoted in Rahaim, "That Ban(e) of Indian Music," 669.

148 Rahaim, "That Ban(e) of Indian Music," 658–59.

149 Rahaim, "That Ban(e) of Indian Music," 672.

150 For example, the former physician Vidyadhar Oke has patented a harmonium based on a twenty-two-*sruti* tuning system, with which he wishes to replace equal temperament in Indian music: http://www.22shruti.com/about_dr_oke.asp.

151 Guettat, "Le mālūf tunisien"; and Ruth F. Davis, "Jews, Music and the Power to be Heard: Charting the Early Tunisian Ughniyya to the Present Day," in *Music and the Play of Power in North Africa, Middle East and Central Asia*, ed. Laudan Nooshin (Farnham: Ashgate, 2009).

152 Jean Lambert, "Retour sur le Congrès du Caire de 1932. Identité et acculturation: les prémisses de la mondialisation," *Actes du Congrès des musiques dans le monde de l'islam (Assilah, 8–13 Aug. 2007)*; and Jihad Racy, "Historical Worldviews of Early Ethnomusicologists: An East-West Encounter in Cairo, 1932," in Philip V. Bohlman and Daniel M. Neuman (eds.), *Ethnomusicology and Modern Music History* (Urbana: University of Illinois Press, 1991).

153 Delegates at the congress were divided into seven committees: 1) General Issues, 2) Melodic Modes, Rhythmic Modes, and Composition, 3) the Musical Scale, 4) Instruments, 5) Recording, 6) Music Education, and 7) Music History and Manuscripts. Racy, "Historical Worldviews of Early Ethnomusicologists," 71.

154 Racy, "Historical Worldviews of Early Ethnomusicologists," 76–77.

155 Alexander J. Ellis, "On the Musical Scales of Various Nations," *Journal of the Society of Arts* 33, no. 1688 (1885): 485–527.

156 Harry Liebersohn, *Music and the New Global Culture: From the Great Exhibitions to the Jazz Age* (Chicago: University of Chicago Press, 2019), 121.

157 Alice C. Fletcher, "Indian Music. An Address Delivered in Washington, D.C., April 1894," *Music: A Monthly Magazine* (June 1894): 192–193.

158 Ogle, *The Global Transformation of Time*, 8.

159 "For the purpose of phonographic recordings a small pitch pipe is blown as prescribed into the recording horn by explorers, in order later to be able to repro-

duce the original speed of the cylinder, and hence the pitch and tempo of the song." Carl Stumpf, *The Origins of Music*, ed. and trans. David Trippett (Oxford: Oxford University Press, 2012), 36. See also: "At the end of the roller . . . a narrow strip remains free. Here, the normal a1 (= 435 v.d.) of a pitch pipe is recorded later—later because experience has shown that the specification of a tone influences the intonation of the singer when recording vocals. By returning to the normal tone, the original speed of the recording can be reproduced with exactitude"; "The other steps in the playback correspond to those in the recording: placing the roller on the cone, completely winding the clockwork, setting the original speed (and pitch) according to the recorded normal tone," Erich Moritz von Hornbostel, "Phonographische Methoden," in *Handbuch der biologischen Arbeitsmethoden, Abt. V, Methoden zum Studium der Funktionen der einzelnen Organe des tierischen Organismus, Teil 7 (1. Hälfte)*, ed. Emil Abderhalden (Berlin: Urban & Schwarzenberg, 1930), 426–427 and 429. The recordings of the Congress of Cairo in 1932 feature similar points of reference.

160 Alexander Rehding, "The Quest for the Origins of Music in Germany Circa 1900," *Journal of the American Musicological Society* 53, no. 2 (2000): 345–385; Eric Ames, "The Sound of Evolution," *Modernism/Modernity* 10, no. 2 (2003): 297–325; Carl Stumpf, *The Origins of Music*, trans. David Trippett (New York: Oxford University Press, 2012); and Ana Petrov, *Rethinking Rationalisation: Evolutionism and Imperialism in Max Weber's Discourse on Music* (Vienna: Hollitzer, 2016).

CHAPTER THREE

Letter to Charles Eliot Norton, 4 Feb. 1872, in *Letters of Henry James*, vol. 1, ed. Leon Edel (Cambridge, MA: Harvard University Press, 1974), 274.

1 Thompson, *The Soundscape of Modernity: Architectural Acoustics and the Culture of Listening in America, 1900–1933* (Cambridge, MA: MIT Press, 2002).

2 Boston's first music hall was inaugurated in 1852.

3 *The Great Organ in the Boston Music Hall: Being a Brief History of the Enterprise from Its Commencement, with a Description of the Instrument; Together with the Inaugural Ode, and Some Account of the Opening Ceremonies on the Evening of November 2, 1863; to which is Appended a Short Account of the Principal Organs in England and on the Continent of Europe* (Boston: Ticknor and Fields, 1865).

4 In the early 1850s, Boston had seen imposing realizations by local builders E. G. G. Hook in Tremont Temple and William B. D. Simmons in Williams Hall. Barbara Owen, *The Great Organ at Methuen* (Richmond, VA: Organ Historical Society Press, 2011), 23. For a view on the national situation at the time, see Orpha Ochse, *The History of the Organ in the United States* (Bloomington: Indiana University Press, 1975).

5 Ochse, *The History of the Organ in the United States*, 200.

6 Bill F. Faucett, "Prologue: Beethoven's Statue," in *Music in Boston: Composers, Events, and Ideas, 1852–1918* (Lanham, MD: Lexington Books, 2016), xix–xxiii.

7 Ochse, *The History of the Organ in the United States*, 197. The program opened with the performance of the Great Toccata in F and the Trio Sonata in E-flat.

8 On the historicist implications of the pitch chosen by the French commission, see above, chapter 1. Well below the standards in use at the time in Europe, it was meant to favor the performance of vocal and early music.

9 *The Great Organ in the Boston Music Hall*, 20.

10 The sociopolitical implications of their actions and consequences on the development of "high" and "low" culture in America are much debated among scholars. Several works have challenged the concept of a "sacralization" of Boston's musical culture in the late 1800s, as developed especially by Lawrence L. Levine, *Highbrow/Lowbrow: The Emergence of Cultural Hierarchy in America* (Cambridge, MA: Harvard University Press, 1988); and Paul DiMaggio, "Cultural Entrepreneurship in Nineteenth-Century Boston: The Creation of an Organisational Base for High Culture in America," *Media, Culture, and Society* 4 (1982): 33–50, arguing that there was far more impurity than Levine claims. Joseph Horowitz, *Classical Music in America: A History of Its Rise and Fall* (New York: W. W. Norton, 2005); and Nancy Newman, *Good Music for a Free People: The Germania Musical Society in Nineteenth-Century America* (Rochester, NY: University of Rochester Press, 2010).

11 Charles C. Perkins and John Sullivan Dwight, *History of the Handel and Haydn Society, of Boston, Massachusetts, from the Foundation of the Society through its Seventy-Fifth Season: 1815–1890* (Boston: Alfred Mudge & Son, 1883–1893).

12 Michael Broyles, *Beethoven in America* (Bloomington: Indiana University Press, 2011), 30–35.

13 Newman, *Good Music for a Free People*.

14 Mary Wallace Davidson, "John Sullivan Dwight and the Harvard Musical Association Orchestra," in *American Orchestras in the Nineteenth Century*, ed. John Spitzer (Chicago: Chicago University Press, 2013).

15 Faucett, *Music in Boston*, 1–3.

16 The results of Upham's study were published in *Records of Maculated Typhus, or Ship Fever: With Suggestions of Treatment: being the result of a series of observations made at South Boston and Deer Island hospitals in 1847–48* (New York: J. F. Trow, Printer, 1852).

17 Upham was the president of the Handel and Haydn Society from 1860 until 1870 and the chairman of the music committee of Boston's public schools between 1857 and 1872.

18 Edward J. Gillin, *The Victorian Palace of Science: Scientific Knowledge and the Building of the Houses of Parliament* (Cambridge: Cambridge University Press, 2017). Upham's articles on acoustics were reprinted in the *American Journal of Art and Science* and published together under the title *Acoustic Architecture of the Construction of Buildings with Reference to Sound and the Best Musical Effect* (New Haven, CT: R. L. Hamlen, 1853). The following notes refer to this edition.

19 Upham, *Acoustic Architecture*, 6.

20 Upham, *Acoustic Architecture*, 28.

21 Upham, *Acoustic Architecture*, 21.

22 Upham, *Acoustic Architecture*, 5.

23 "The most sublime effects of the oratorio and symphony can only be produced in spacious buildings" (Upham, *Acoustic Architecture*, 26). Upham's views on architecture and taste for large-scale musical venues appear in the impressions he published of his European tour in 1853 (see below). About the Salle des Menus-Plaisirs, home of the Paris Conservatory's orchestra which served as a museum for classical music, he wrote: "Why is it, I could not but ask myself, that such an orchestra . . . be cribbed and cabined in so limited a sphere? This question applies with equal fitness to most of the cities of Germany. Really, there is not on the continent of Europe, so far as I could learn, a concert room of sufficient amplitude to give orchestral music its proper and legitimate effect." Upham, "Reminiscences of a Summer Tour," *Dwight's Journal of Music* 9, no. 5 (3 May 1856): 35.

24 Upham, *Acoustic Architecture*, 35 and 40.

25 Oliver Wendell Holmes, "The Great Instrument," *Atlantic Monthly* (Nov. 1863): 644. The article goes on to state that in the Civil War, during which Upham served as a surgeon, he also developed a "system of sound-signals . . . applied with the most promising results, as a means of intercommunication between different portions of the same armament."

26 From the end of May through 13 June, Upham observed typhus fever in London. His goal was to compare the disease "in its *indigenous haunts*, with its manifestations and habits here [in Boston] as *an exotic*." Upham, *Illustrations of Typhus Fever in Great Britain: The result of Personal Observations made in the Summer of 1853 with some Remarks as to its Origins, Habits, Symptoms and Pathology; to Which is Appended a Brief Account of the Re-Appearance of Typhus in Boston in the Winter of 1857–58* (Boston: David Clapp, 1854), 6.

27 Several articles on the organ present Upham's travel of 1853 as primarily dedicated to the visit of European organs, as does Upham's own narration of this tour, anonymously published in *Dwight's Journal of Music* under the title "Reminiscences of a Summer Tour."

28 J. B. Upham, *Narration of My Organ Tour: Remarks at a meeting of the Harvard Musical Association*, Harvard Musical Association, January 1858. I thank Jim McDonald for sharing a copy of this document with me.

29 "Classical music in the United States is a mutant transplant . . . . The resulting foliage, oftentimes resplendent, was as often peculiar." Horowitz, *Classical Music in America*, xiii.

30 Owen, *The Great Organ*, 79.

31 Owen, *The Great Organ*, 69 and 74.

32 "Christmas Oratorio" (9 Jan. 1864): 167.

33 A local organist said that the organ "spoke day after tomorrow," while some of his colleagues pretended they had to start playing an hour or even a day ahead of the concerts in order to be synchronized with the orchestra and the choir. Owen, *The Great Organ*, 75.

34 [Geo. H. Chickering], "Annual Meeting of the Handel and Haydn Society," *Dwight's Journal of Music* (8 July 1865), 59. The purchase included six clarinets, two flutes, two oboes, and two bassoons for a total of $300 on the Handel and Haydn Society's budget.

35 "Oratorio," *Dwight's Journal of Music* (7 Dec. 1867): 151. The review refers to an "unusually large orchestra," which might indicate the addition of musicians and instruments who had never played with the Society before.

36 Such is the scenario suggested by Charles R. Cross, author of the first accounts on the history of pitch in America. Charles R. Cross, "Historical Notes Relating to Musical Pitch in the United States," *Proceedings of the American Academy of Arts of Sciences* 35 (1900): 453–454.

37 Owen's suggestion that "no one thought of negotiating the pitch" thus seems highly unlikely, especially as Hopkins served as an intermediary in the negotiations (he published a book containing detailed praise of the low pitch, reviewed at length in *Dwight's Journal*: "Musical Pitch. Extracted from 'The Organ and its History and Construction.' By E. J. Hopkins and E. F. Rimbault. LL.D.," *Dwight's Journal of Music* 23, no. 14 [3 Oct. 1863]: 108–9).

38 The first of many reports on British conversations was published in June 1860: "Musical Pitch (From the London Athenaum)," *Dwight's Journal of Music* 17, no. 13 (23 June 1860): 101.

39 Edward Wiebe, "An American Standard Diapason" (21 Jan. 1865): 379–809.

40 *Dwight's Journal of Music* 23, no. 14 (3 Oct. 1863): 110–111.

41 "In Boston, the Music Hall is wisely pitched to the French normal diapason, but our orchestral pitch is about three-eights of a tone higher, or 445 vibrations.... This is more than a tone higher than a hundred years ago, when the human voice regulated the musical pitch, as it always should do. It is a great error of judgement to suppose that the orchestral instruments, or the organ and piano, sound better at such a high pitch. They may gain a certain brilliancy, but they lose, on the other hand, a good deal of their peculiar quality of resonance of tone." *Dwight's Journal of Music* 30, no. 25 (25 Feb. 1871): 403.

42 Paine Hall's ceilings serve as the opening example in Katherine Bergeron and Philip V. Bohlman (eds.), *Disciplining Music: Musicology and Its Canons* (Chicago: University of Chicago Press, 1992), 1.

43 *Dwight's Journal of Music* 29, no. 3 (24 April 1869): 23 (quoted in Owen, *The Great Organ*, 116).

44 "During the month of May, there was much thought and discussion in our city, as well as in musical circles everywhere, on the important question of a reform of the concert pitch." Dwight, *History of the Handel and Haydn Society*, 283.

45 "Concerts. Normal Diapason," *Dwight's Journal of Music* 29, no. 6 (5 June 1869): 47. In the later version of this report, Dwight corrected his mistake and wrote "(down)" in place of "up." Dwight, *History of the Handel and Haydn Society*, 284. The program was formed by Beethoven's Eighth Symphony, an aria by Mozart, an overture by Cherubini, and a hymn by Mendelssohn.

46 Dwight, *History of the Handel and Haydn Society*, 284.

47 Dwight, *History of the Handel and Haydn Society*, 289.

48 Dwight, *History of the Handel and Haydn Society*, 295.

49 Owen, *The Great Organ*, 189.

50 Sandy Raymond Mazzola, "When Music Is Labor: Chicago Bands and Orchestras and the Origins of the Chicago Federation of Musicians, 1880–1902" (PhD diss., Northern Illinois University, 1984), 297.

51 The organization was created in 1876 by Eben Tourjée, founder of the New England Conservatory. The creation of this music school inspired by European conservatories was the result of a collaboration among Tourjée, Dwight, and Upham. Bruce MacPherson and James Klein, *Measure by Measure: A History of New England Conservatory from 1867* (Boston: Trustees of New England Conservatory of Music, 1995).

52 American Federation of Musicians, *Official Proceedings of the Seventh Annual Convention Held at Lyric Hall, Buffalo, New York, June 3, 4, 5, 6 and 7, 1902* (St. Louis, MO: Allied Printing Trades Council), 76.

53 "Col. Moore on Musical Pitch." The Piano Manufacturers' Association of New York was the first piano makers' organization in the country. It was created in fall 1890, and pitch standardization was thus one of its first projects.

54 The first mention of conversations on standard pitch among members of the Piano Manufacturers' Association in William Steinway's diary dates from 10 March 1891: "at 3 P.M. large attendance of Piano Manufacturers Assoc. . . . discuss Standard pitch." On 31 March, Steinway noted: "At 5 P.M. at Meeting and Reception of Piano M[anu]f[ac]t[u]r[e]rs. and their friends, make speech on standard pitch." Held by the National Museum of American History, the William Steinway Diary 1861–1896 has been digitized and is available at http://americanhistory.si.edu/steinwaydiary/.

55 H. Paul Mehlin (G. Mehlin & Sons, New York), Ernest Knabe (Wm. Knabe & Co., Baltimore), William T. Miller, Thomas Scanlan (New England Piano Company, Boston), and C. E. Elsbree (Lester Piano Company, Philadelphia).

56 "The committee appointed by the piano manufacturers . . . beg leave to report that they . . . have prosecuted the work both in this country and in Europe . . . . Sufficient evidence has been collected to not only show the importance of a uniform pitch, but of the great benefits to be derived from the adoption by the piano manufacturers of such a standard pitch as had commended itself to the largest of musical people. Your committee therefore recommend the adoption as a standard musical pitch that A which gives 435 double vibrations in a second of time." "Standard Pitch Adopted by the Piano Manufacturers' Association of New York and Vicinity, Friday, November 6, '91. Fuller's Address," *The Musical Courier* 23, no. 612 (11 Nov. 1891): 549.

57 At the time of his work for the committee, Fuller was vice president of the Estey Organ Company. Fuller built his own observatory and library and collected hundreds of forks, which played an important role in the history of pitch standardization in the United States. Following his 1891–1892 campaign in favor of the French pitch, Fuller became the forty-fourth governor of Vermont (from 1892 to 1894). Dennis G. Waring, *Manufacturing the Muse: Estey Organs and Consumer Culture in Victorian America* (Middletown, CT: Wesleyan University Press, 2002).

58 Born in Troy, New York, in 1848, Cross studied at MIT, where he became an instructor in physics after his graduation in 1870. A disciple of MIT's celebrated mathematician Edward C. Pickering, Cross belonged to an early generation of students trained according to the new principles of laboratory culture developed at the Institute since 1865.

59 Letter from L. K. Fuller to Charles R. Cross, Brattleboro, 22 March 1883. Research Correspondence 1883–1898, Charles R. Cross Papers, Massachusetts Institute of Technology, Institute Archives and Special Collections (hereafter Cross papers).

60 On the relationship between Bell and Cross, see Robert V. Bruce, *Bell: Alexander Graham Bell and the Conquest of Solitude* (Ithaca, NY: Cornell University Press, 1973), 110–112.

61 "Deposition of Charles R. Cross," in United States Circuit Court, Southern District of New York. *American Bell Co. et al. vs. the People's Telephone Co. et al.* Evidence for Complainants, Vol. 1.—Evidence (Boston: Alfred Mudge and Sons, 1882), 20–51.

62 Pantalony, *Altered Sensations*, 76.

63 Cross conducted several experiments on the sound of vowels with the phonograph: "Helmholtz's Vowel Theory and the Phonograph," *Nature* 18 (23 March 1878): 93–94; and Cross and George V. Wendell, "On Some Experiments with the Phonograph, Relating to the Vowel Theory of Helmholtz," *Proceedings of the American Academy of Arts and Sciences* 27 (1891–1892): 271–279. He also performed experiments on pitch discrimination: Cross, "On a Proposed Method for Ascertaining the Least Number of Vibrations Necessary to Determine Pitch," *Proceedings of the American Association for the Advancement of Science* 33 (1884): 113–114; and Cross and Margaret Eliza Maltby, "On the Least Number of Vibrations Necessary to Determine Pitch," *Proceedings of the Academy of Arts and Sciences* 27 (1893): 222–235.

64 William T. Miller, who later took over his father's business, was a member of the Piano Manufacturers' Association's Pitch Committee in 1891.

65 Cross and William T. Miller, "On the Present Condition of Musical Pitch in Boston and Vicinity," *American Journal of Otology* 2, no. 4 (1880): 249–263.

66 Cross and Miller's article contains a mention of Koenig's most recent paper, published just a few months before in the *Annalen der Physik*: "Untersuchungen über die Schwingungen einer Normalstimmgabel," *Annalen der Physik* 9 (1880): 394–417. Pointing out the great influence of temperature on the small tuning devices used by makers, Cross and Miller explained how, to guarantee objective observation and drawing on Koenig's observations, they used cork to insulate the forks from their bodies' warmth. On Koenig's work on temperature, see Pantalony, *Altered Sensations*, 91–92 and 102–104. Cross and Miller used an "O fork of [Koenig's] vowel-series," which Cross may have employed when working on vowels' tones ("On the Present Condition of Musical Pitch in Boston," 251). Cross and Miller's experiment also benefited from the connection between Koenig and the American physicist Alfred Marshall Mayer, the supplier of the etalon instrument that played a central part in their experimental arrangement. Cross and Miller, "On the Present Condition of Musical Pitch in Boston," 249. Mayer (1836–1897) left the United States to study in Paris in 1863. He was "one of Koenig's closest collaborators and supporters" (Pantalony, *Altered Sensations*, 111). He founded the Stevens Institute of Technology in Hoboken, New Jersey.

67 The specific qualities of tuning forks that makers employed (their small size, in particular), though challenging the established protocols of physics laboratories, especially provided the two experimentalists with an opportunity to show off their knowledge of these sound objects. In order to overcome the "considerable difficulty . . . experienced on account of the brief duration of the sound" these

small forks produced, Cross and Miller extensively repeated their observations and comparisons. Cross and Miller, "On the Present Condition of Musical Pitch in Boston," 252.

68 While in his 1880 paper Ellis mentioned only one American pitch, measured from a fork "obtained from New York by Messrs. Steinway," in 1885 he included Cross's data within the tables of the second edition of his Helmholtz translation, giving the United States, and especially Boston, a prominent place in debates regarding pitch. "Appendix," in Helmholtz, *On the Sensations of Tone*, 511.

69 This method was first introduced by Heinrich Scheibler in *Der physikalische und musikalische Tonmesser* (Essen: G. D. Bädeker, 1834). See Jackson, *Harmonious Triads*, 151–181 and 206–207.

70 The collection includes four forks by Parisian maker Marloye, which were imported by the Harvard professor of natural sciences Joseph Lovering (they are now part of the Harvard Collection of Historical Scientific Instruments). Three of these forks were supposed to give the pitch in use at Vienna, Florence, and Milan in the 1840s. The collection also includes the "Covent Garden" pitch, embodied by a fork belonging to a Brooklyn builder.

71 As Cross would later put it, the United States "followed the usage [from] abroad." "Historical Notes Relating to Musical Pitch in the United States," *Proceedings of the American Academy of Arts of Sciences* 35 (1900): 453.

72 "The history of the pitch of the great organ is, however, so well known, that it need only to be referred to here." Cross and Miller, "On the Present Condition of Musical Pitch in Boston," 260.

73 Cross and Miller, "On the Present Condition of Musical Pitch in Boston." This narrative became firmer in the article Cross published about the history of pitch in the United States: "The great harm arising from the excessive height to which the pitch had risen . . . was recognized by those interested in the procurement of the Great Organ for the Boston Music Hall." Cross, "Historical Notes," 453.

74 In 1900, a few years after Fuller's death, Cross presented the results of his decades-long experiments in a substantial article, which still serves as one of the main sources on the history of pitch in the United States. Cross, "Historical Notes," 453–454.

75 This becomes clear in Fuller's manner of addressing the physicist as a consultant. In the spring of 1891, for instance, as he was preparing to address the Piano Manufacturers' Association, Fuller wrote to Cross requesting information on the relationship between *A* and *C* standards in equal temperament and "true" intonation. As Fuller put it, this was a delicate question "arising now so frequently I thought it wise to ask a better man, and I feel the strength which comes from your statement. You have made it very plain and I can go ahead with greater confidence." Letter from Fuller to Cross, 5 May 1891. Cross papers.

76 Fuller claimed he sent 1,000 copies of his first Circular, explaining the problem and calling for opinions on the matter, and forks representing the pitch in use in factories or musical institutions. Letter from Fuller to Steinway, 13 May 1891. Fuller's papers. Letters, no. 18. "Uniform pitch" (hereafter Fuller's papers).

77 Circular no. 5 was a copy of the English translation of the 1859 French commission's report (transmitted by the Journal of the Society of Arts); no. 6 gave an account of the Vienna Congress and other European decisions; no. 7 was a reprint

of Fuller's review of Koenig's research; and no. 8 a diagram displaying the number of vibrations for the notes of a tempered scale based on *A* 435. Fuller's papers.

78 Fuller's correspondents included the successive presidents of the National League of Musicians, the heads of the New York and Boston philharmonic societies, the president of the National Teachers' Association, and the head of the Cincinnati Conservatory of Music. Fuller's papers.

79 Letter from Fuller to Steinway, 26 April 1891. Fuller's papers.

80 Letter from Fuller to Sherman, Clay & Co, 26 Sept. 1891. Fuller's papers.

81 Letter from Fuller to Cross, Brattleboro, Vermont, 15 Sept. 1891. Cross papers.

82 Frank D. Abbott, *Musical Instruments at the World's Columbian Exposition* (Chicago: Presto Co., 1895), 174.

83 See chapter 1.

84 "Standard pitch adopted by the Piano Manufacturers' Association," 549.

85 "Standard pitch adopted by the Piano Manufacturers' Association," 549.

86 Letter from Fuller to Mason & Hamlin Organ & Piano Co., 22 June 1891. Fuller's papers. Fuller reached Southampton on 6 June 6 1891 and visited Bremen, Hamburg, Paris, London, Edinburgh, and Liverpool. Untitled document. Fuller's papers. Box 4, "Itineraries."

87 "After spending a day with Koenig, I am thoroughly convinced as to the accuracy of his system and methods, and in addition, Mr. Hipkins of London, Sir Henry Wood of the Society of Arts, Mr. Blakely, the late Mr. Ellis, Mr. Webb, and other prominent men of Great Britain have assured me that there is no one in Europe with the reputation of Dr. K-." Letter from Fuller to Cross, Brattleboro, 8 Aug. 1891. Cross papers.

88 "Mr. Ellis and Mr. Hipkins from London, have given the matter of Valentine & Carr's forks personal attention, and, I believe, they will give a better tone with longer duration than the cheaper German ones." Letter from Fuller to J. W. Queen & Co., 6 Aug. 1891. Fuller's papers.

89 Letter from Fuller to J. H. Carr, 23 Jan. 1892. Fuller's papers.

90 Letter from Fuller to Carr, 23 Jan. 1892. Fuller's papers.

91 Letter from Fuller to Daniel Spillane, 24 May 1891. Fuller's papers.

92 Letter from Fuller to Carr, s.d. [22 or 23 Sept. 1891]. Fuller's papers.

93 The Association had to await the decision of an attorney regarding the design of the fork. As one had to let the fork rest after it had been worked on, this added a significant delay to the process.

94 It was feared that when the forks would reach the ports of Boston and New York, they would have to wait another few weeks. But Fuller went around this problem by writing to the port authorities.

95 Valentine and Carr did not conform to the orders sent by Fuller, who wanted to materialize the difference between *A* and *C* standards by using different colors for the cases.

96 Letter from Fuller to Valentine and Carr, 31 Dec. 1891; letter from Fuller to Steinway, 2 Jan. 1891. Fuller's papers.

97 Letter from Fuller to J. H. Carr, 23 Jan. 1892. Fuller's papers.

98 The main criteria in selecting these candidates was the size of their business: they should be able to flood the country with the standard. As elsewhere, however, dealers authorized by the Piano Manufacturers' Association suffered from the competition of retailers selling "cheap German forks" of less accuracy, as well as from the public's taste for mouthpipes. See Letter from Fuller to William E. Wheelock, 15 Feb. 1892. Fuller's papers.

99 Letter from Fuller to J. C. Haines & Co., 4 Dec. 1891. Fuller's papers.

100 Pantalony, *Altered Sensations*, 91–92.

101 "The object of the present invention is to produce tuning-forks of uniform size and absolutely-correct pitch and tone with the minimum of material and labor. The new method of production consists in forming the handle of the fork of a separate piece of steel from the prongs of the fork and then uniting the handle and fork-prongs in such a manner as to render the handle homogeneous with the fork-prongs, with no break in the continuity of the metal, the union of the handle and the fork-prongs being so effected that the subsequent finishing of the fork and bringing it in correct tune are facilitated." Fuller, "Method of Making Tuning-Forks," US Patent Office, Patent Number 483,513, 27 Sept. 1892, 1.

102 Fuller's letters to Cross contain evidence of this resistance from a number of firms. In a letter dated 5 May 1891, Fuller reported that in response to his questionnaire, he had received the following answer: "Taking French Pitch A 435 as the lowest / And N.Y. Orchestral High A 460 as the highest / Subtract the difference 25 / Divide the result by 2 12,5 / Add it to the Diapason Normal A 435 and it will produce / the happy Medium Am[erican]. Nat[ional]. Pitch A 447,5."

103 Joshua D. Gailey, "Beginning Bands: Progressive Reform and the Birth of the American School-Band Industry, 1907–1940" (PhD diss., Yale University, 2019), 140n39.

104 In 1907, for instance, in the Federation's convention proceedings, one reads: "There are some theatrical circuits still using High pitch, which fact makes conditions very unsatisfactory." The following year, the executive board of the Federation resolved "that a committee be appointed by the President, whose duty it shall be to determine and carry out some line of action whereby we can secure the cooperation of the different vaudeville circuits and booking agents, tending to the adoption and exclusive use of International Pitch throughout the country in all of the different theaters owned and controlled by the different managements and agents, as mentioned above." *Official Proceedings of the Eighth Annual Convention of the American Federation of Musicians, Held at Cleveland, Ohio, May 20, 21, 22, 23, 24, 25, 1907* (St. Louis, MO: Allied Printing Trades Council), 157; and *Official Proceedings of the Ninth Annual Convention of the American Federation of Musicians, Held at Aschenbroedel Hall, 3535 Pine Street, Saint Louis, Missouri, May 11 to 16, 1908*, 161.

105 "French Musical Pitch Chosen. For Bands and Other Musical Instruments at the World's Fair," *The Music Trade Review* 36, no. 7 (1903): 19.

106 "Some Discussions About 'Pitch.' Director Stewart Insists that All Bands Competing at St. Louis Shall Play in the International Pitch—Some Varied Views on the Subject," *The Music Trade Review* 28, no. 12 (1904): 9.

107 For syntheses on American imperialism at the turn of the century, see Joseph A. Fry and Adam Burns, *American Imperialism: The Territorial Expansion of the*

United States, 1783–2013 (Edinburgh: Edinburgh University Press, 2017); Antony Gerald Hopkins, *American Empire: A Global History* (Princeton, NJ: Princeton University Press, 2019), 239–436; and Thomas Bender, *A Nation Among Nations: America's Place in World History* (New York: Hill and Wang, 2006), 182–245.

108 Anti-German feelings arose well before the war in the musical field, as an expression of nationalism and a reaction to Germany's cultural domination throughout the second half of the nineteenth century. Jessica Gienow-Hecht, *Sound Diplomacy: Music and Emotions in Transatlantic Relations, 1850–1920* (Chicago: University of Chicago Press, 2009), 151–176.

109 "A Musical Innovation of Importance to American Orchestras," *The Sun*, 13 Feb. 1919, 6. *A* 440 was first suggested as a standard during a conference of natural scientists held in Stuttgart in 1834. See Jackson, *Harmonious Triads*, 183–230.

110 William Braid White, "Our Technical Department," *The Music Trade Review* 66, no. 25 (22 June 1918): 16.

111 Roland Wittje, *The Age of Electro-Acoustics: Transforming Science and Sound* (Cambridge, MA: MIT Press, 2016); and Thompson, *The Soundscape of Modernity*.

112 William Moersch, "The Marimba Revolution: Mallet Instruments, Repertoire, and Technique in the Twenty-First Century," in *The Cambridge Companion to Percussion*, ed. Russell Hartenberger (Cambridge: Cambridge University Press, 2016), 43–44.

113 Documenting Deagan's activities is challenging, given that most of his archives were destroyed in a fire. Instead, for a historical account of his career, we must rely on what is often unreliable information from the corporate documents of the Deagan Company. According to the literature produced by Deagan Inc., for instance, the adoption of *A* 440 by the American Federation of Musicians is supposed to date from 1910, which goes against all other evidence on the matter.

114 On the entanglement of acoustical science and instrument building in the context of industrialization, see Sonja Petersen, "Craftsmen-Turned-Scientists: The Circulation of Explicit and Working Knowledge in Musical-Instrument Making 1880–1960," in Hui, Kursell, and Jackson, "Music, Sound and the Laboratory," 212–223.

115 Tim Gracyk and Frank Hoffmann, *Popular American Recording Pioneers: 1895–1925* (London: Routledge, 2008).

116 The first sound cartoons were released in 1928, and include shorts produced by various studios before Disney's *Steamboat Willie*, often mistakenly credited as the first one. James Bohn, *Music in Disney's Animated Features:* Snow White and the Seven Dwarfs *to* The Jungle Book (Jackson: University Press of Mississippi, 2017).

117 Deagan provided the channel NBC with its long-lived sound signature, a C-major chord played by three chime-bars; see http://www.nbcchimes.info/dnbc.php. On the entanglement of pitch standardization and broadcasting, see below, chapter 4.

118 Fred Dahlinger Jr., "Ringing and Ringlings: Showmens' Bells, Chimes and Related Novelty Musical Instruments—Part II," *Carousel Organ*, no. 20 (2004): 16. I have not yet been able to verify Deagan's colorful claims regarding his extensive travels across the world: "The writer, Mr. J. C. Deagan, an original investigator and thinker, has devoted a lifetime of study to the science and philosophy of

musical tones. He takes nothing for granted, knowing that history and tradition were not trustworthy, and with a burning desire to know the highways and byways by which music has come down to us, has made a personal investigation in all corners of the earth for facts, more facts, concerning the origins of musical instruments, scales, pitch, etc. A number of years ago he studied in Europe under the great Helmholtz. In research work he has visited the leading Universities, Museums, Literary and Musical Circles of this country and Europe. In Greece he examined the ancient musical instruments, such as the Lyras referred to in this article, some of them having been excavated from the great temples of Apollo at Argos, Delos and Delphi. Mr. Deagan has also visited Egypt many times, has gone a thousand miles into the Sahara Desert, where he found the customs, habits, manners, morals and music the same as in the days of the ancient Pharaos (nothing changes in the desert). He examined many musical instruments belonging to the ancient world, some of which were in a wonderful state of preservation, having been taken from Temples and Tombs that had been covered and preserved by the sands of the Sahara for thousands of years." J. C. Deagan, "Editorial Note," *The Dea-Gan-Ometer. Patented, Other Patents Pending. A New Instrument for Easily Ascertaining and Standardizing any Desired Pitch on a Scientific Basis* (Chicago: J. C. Deagan, 1916), 2.

119 "J. C. Deagan . . . is acknowledged to be one of the greatest living authorities on matters pertaining to the science of acoustics, pitch and tone production and in all of the instruments produced in our immense fireproof factory are incorporated the results of years of experimenting and research on the part of a corps of skillful artists working under Mr. Deagan's personal supervision. . . . [Deagan] brought [the xylophone] out of the toydom . . . . The South African Marimba, which was simply a series of war drums, was next developed into the forms which the civilized world knows today." J. C. Deagan Inc., "Introductory," *Catalog R* (1920), Percussive Arts Society, courtesy James A. Strain, 2–3.

120 Deagan Inc., *Catalog R*, 22.

121 *The Dea-Gan-Ometer*, 3.

122 Jackson, *Harmonious Triads*, 151–181. Deagan also commercialized a more common tonometer consisting of thirteen tuning forks.

123 On this technology, see above, chapter 2.

124 *The Dea-Gan-Ometer* also referenced Scheibler's work through Deagan's selection of pitches for the collection of bars forming the tuning device, which included *A* 435 and *A* 440, as well as all the frequencies in between.

125 It is unclear how many of these documents were printed and what their intended use was. The texts were advertised in Deagan's catalogs, where one reads that they could be sent for free to any interested customer. Deagan also allegedly issued all of these prospectuses in the form of a single book, though I have not yet been able to locate this.

126 John C. Deagan, *Fundamentals in A=440 Pitch. Harmonic Intervals at 22°C (72°F)* (Chicago: J. C. Deagan, 1916). Library of Congress / A478761.

127 Helmholtz was opposed to equal temperament, which he saw as a distortion of the natural laws of acoustics.

128 Deagan's use of Helmholtz for commercial purposes was also attested in 1927, when the firm employed the title of the German's eminent scientific treatise

to advertise one of its latest inventions and most successful achievements, the "Vibraharp," which would be developed into what twentieth-century audiences would know as the vibraphone. Deagan declared this device to be "A New Sensation in Musical Tone," deploying the English title of Helmholtz's magnum opus commercially to gain an advantage over his competitor Leedy, who had started developing vibraharps in 1922.

129 Deagan, *Fundamentals—A Ready Reference in A=440 Pitch Calculations. Musical Mathematics, simplified from the Purely Theoretical to the More Practical, in the Newly Authorized A=440* [stamped 13 Dec. 1917]. Library of Congress / A478761.

130 Deagan, "A=440 Pitch Adopted: Pitch versus Temperature," *Musical Quarterly* 4, no. 4 (1918): 588. Hipkins put the point in very similar terms: "We may regret that the French Commission did not fix their grade of temperature at 20° instead of 15°. They could then have advantageously fixed the normal diapason at *A* 440." A. J. Hipkins, "The Standard of Musical Pitch," *Journal of the Society of Arts* (28 Feb. 1896): 34.

131 Deagan, "A=440 Pitch Adopted," 588.

132 Deagan, "A=440 Pitch Adopted," 592.

133 In Deagan's catalogs, the expression designates both a set of tuning forks and the Dea-gan-ometer.

134 Letter from J. C. Deagan to S. W. Stratton, 20 Aug. 1918. General Records of J. Howard Dallinger, Box 13, National Archives RG 167. Records of the National Institute of Standards and Technology / National Bureau of Standards (Department of Commerce).

135 The Piano Manufacturers' Association comprised several organizations: the National Association of Music Merchants; the Organ Builders' Association of America: the National Association of Music Roll Manufacturers; the National Musical Merchandise Assn. of the U.S.; the Musical Supply Association of America; the Music Publishers' Association of the U.S.; the Band Instrument Manufacturers' Association; the National Association of Talking Machine Jobbers; the National Piano Travelers' Association; and the National Association of Piano Tuners. Herbert Wilder Greene and B. H. Goldsmith (eds.), *Musical Blue Book of America, 1921–2* (New York and Philadelphia: The Musical Blue Book Corporation, 1922), 7–8.

136 Greene and Goldsmith (eds.), *Musical Blue Book of America, 1921–2*, 6–7.

137 Miller's activities as pitch controller in the 1910s have not left many traces. But in 1921, he claimed that he had checked the standard forks used by Lyon & Healy, the Aeolian Co., Steinway, Chickering, Mason & Hamlin, Deagan, Gaertner, and the Standard Scientific Co. White, "Further Thoughts on Pitch. Some Arguments Over the Relative Merits of A-435 and A-440," *The Music Trade Review* 72, no. 22 (28 May 1921): 39.

138 "Chamber appoints Committee on Standard Musical Pitch. Committee to Act in Accordance With Resolution Adopted at Last Directors' Meeting to Establish a Real Standard Pitch," *The Music Trade Review* 80, no. 3 (27 Jan. 1925): 21.

139 "Chamber appoints Committee on Standard Musical Pitch."

140 Here is the detail of the questionnaire: "Question 1. How many chippings are necessary, Question 2. How should we tune in relation to pitch in the course of

the first chipping? Question 3. How much time should elapse between first and second chipping? Question 4. How should we tune in relation to pitch in the course of the second chipping? Question 5. In case a third chipping is necessary, how much time should elapse between second and third chipping? Question 6. How should we tune in relation to pitch in the course of the second chipping? Question 7. How much time should elapse between third chipping and first tuning? Question 8. How should we tune in relation to pitch in the course of rough tuning? Question 9. How much time should elapse between rough and second tuning? Question 10. How should we tune in relation to pitch in the course of second tuning? Question 11. How much time should elapse between second and third tuning? Question 12. How should we tune in relation to pitch in the course of third tuning? Question 13. How much time should elapse between third and fourth tuning? Question 14. How should we tune in relation to pitch in the course of fourth tuning? Question 15. What pitch do you recommend?" White, "Technical Department," *The Music Trade Review* 80, no. 16 (18 April 1925): 16–17.

141 "E. R. Jacobson Heads the Chamber of Commerce for Coming Year. Other Officers Are First Vice-President, E. H. Droop; Second Vice-President, C. D. Greenleaf; Treasurer, Herman Irion; and Secretary and General Manager, Alfred L. Smith—Chamber Meeting Adopts Report on Pitch—Next Convention in New York First Week of June," *The Music Trade Review* 80, no. 24 (13 June 1925): 39.

142 "E. R. Jacobson Heads the Chamber of Commerce."

143 "Western Comment," *The Music Trade Review* 80, no. 25 (20 June 1925): 30.

144 White, "Technical Department," *The Music Trade Review* 80, no. 5 (31 Jan. 1925): 20.

145 See above, chapter 2.

146 https://physics.case.edu/about/history/antique-physics-instruments/phonodeik-2/. Miller collected the press reviews of his demonstrations. See Case Western University Archives. Papers of Dayton Clarence Miller (hereafter Miller's papers), Box 4, "Clippings." On Miller's activities, see William Fickinger, *Miller's Waves. An Informal Scientific Biography* (Cleveland: Case Western University Press, 2011).

147 "Great strides should be made in osculautology during the next year. Not only can the canned kisses of loved ones be preserved, but by means of an instrument known as the phonodeik, the invention of Professor Dayton Clarence Miller, of Case School, which records sound waves with instantaneous photography, the sound of the kisses as well. By the same means the words inevitably accompanying a kiss: 'I love you,' or 'You are the only girl (or boy) I ever loved,' may be photographed and interred with the little dab of rouge—all that is mortal of a kiss. Then some night when you are fat and forty, or old and gray, you may put the record on the phonograph, and whilst your dimming eyes gaze at the love records of your life, hear the familiar cadences of some love voice saying 'them fatal words' again. Pretty sentiment, right?" Edward Griffith, "Kanning Kisses or Immortalizing Mush Stuff," *The Cleveland Leader* (1917). Miller's papers, Box 3, "Newspapers clipping collected by Miller. Vol. 3. Sound."

148 Miller's collection of flutes is now held by the Library of Congress in Washington, DC. Some of his compositions (for flute, flute and piano, and voice and piano), can be found in Miller's papers, Box 4.

149 See for instance "Certificate of Accuracy of a Tuning Fork Submitted by the National Association of Piano Tuners, Inc., of Chicago. Cleveland, Ohio, July 1, 1927." Miller's papers, Box 6, "Tuning forks ratings and varia."

150 White, "Standard 440 A Forks to be Made by Miller. Master Instruments to Be Deposited With Associations and Certified Duplicates to Be Available to the Industry," *The Music Trade Review* 82, no. 6 (6 Feb. 1926): 48.

151 White, "The Technical Department. The Standard of Pitch," *The Music Trade Review* 79, no. 4 (26 June 1924): 34.

152 See for example Aashish Velkar, *Markets and Measurements in Nineteenth-Century Britain* (Cambridge: Cambridge University Press, 2012).

153 Yates and Murphy, *Engineering Rules*; Lampland and Starr, *Standards and Their Stories*; Brunsson and Jacobson, *A World of Standards*. On the relation between standards and organizations, see especially Brunnsson and Rasche, "The Dynamics of Standardization."

154 Gienow-Hecht, *Sound Diplomacy*, 159.

CHAPTER FOUR

"Il arrive que le technicien soit d'esprit plus international que l'homme d'État." *L'esprit de Genève* (Paris: Grasset, 1929), 170.

1 The title of this chapter refers to prime minister Neville Chamberlain's 30 September 1938 declaration according to which he had negotiated "peace in our time" with Germany during the Munich conference.

2 BBC Internal Circulating Memo, 5 Oct. 1937. BBC Written Archives Center (hereafter BBC), E2/437/1. Normal Pitch Standardisation 1937–1940 (Foreign general).

3 Handwritten mention from 13 Oct. 1937. BBC, E2/437/1.

4 British Standards Institution (hereafter BSI), CF (B) 2596. Notes of a Meeting of ISA Sub-Committee No. $43/3^{B}$, Concert Pitch, Held at Broadcasting House, London, On Thursday and Friday, 11 and 12 May 1939. UCLA Special Collections. Knudsen papers. Coll. No 1153. Box 25, folder 4 (hereafter Knudsen papers).

5 Britain was represented by the BSI, France by the Agence Française de Normalisation (AFNOR), Germany by the Deutscher Normenausschuß, and Holland by the Dutch Acoustics Committee.

6 The meeting was presided over by Duval (president of the Comité électrotechnique français), Brylinski, president of the Association française de normalisation, Huber-Ruf, secretary of the International Federation of Standards Associations (ISA), and Le Maistre, general secretary of the Commission électrotechnique internationale (CEI). *Journal des télécommunications* 4, no. 11 (Nov. 1937): 295–297.

7 Wittje, *The Age of Electro-Acoustics*. On interwar acoustic knowledge and industry, see also Thompson, *The Soundscape of Modernity*; Bijsterveld, *Mechanical Sounds*; and Mills, "Deafening."

8 BSI CE (B) 7049. Technical Committee B/26, Acoustics, *Translation of Minutes of International Conference on Acoustics, Paris June 30–July 3, 1937*, 23 Feb. 1938;

ISA 43/0-GS 2582, *Compte Rendu de la Réunion de Paris de Juillet 1937*, 29 April 1938. Knudsen papers.

9 Subcommittee 43/3[B] was created within ISA's subcommittee 43/3, on electro-acoustics and music, in the wake of the 1937 conference. Chaired by the German physicist Martin Grützmacher, subcommittee 43/3 comprised 25 delegates from Germany, Austria, the United States, France, Britain, Italy, Norway, Holland, Romania, Switzerland, Czechoslovakia, as well as representatives of the International Electrotechnical Commission, the International Telephone Consultative Committee, the ISA, and the International Broadcasting Union (IBU). "For economy, much of the technical work [of ISA] would be farmed out to the different national bodies which would thus serve as informal secretariats." Yates and Murphy, *Engineering Rules*, 117.

10 Letter from the German secretariat to the members of subcommittee 3. Electro-acoustics, Musical Acoustics of ISA 43. Acoustics, and report on the measurement of the concert pitch a' in musical broadcasts, submitted by Werner Lottermoser, Phys.-Techn. Reichsanstalt, Berlin, Berlin, 12 Dec. 1937. Surrey History Centre. Broadwood Piano Manufacture (hereafter Broadwood papers), 2185/JB/71/4a.

11 See also Edward Pawley, "The Problem of Standardising Musical Pitch Is on the Fringe of BBC Engineering," *BBC Engineering, 1922–1972* (London: British Broadcasting Corporation, 1972), 124.

12 Christiane Sibille, *Harmony Must Dominate the World: Internationale Organisationen und Musik in der ersten Hälfte des 20. Jahrhunderts* (Bern: Quaderni di Dodis, 6, 2016). On the League's attempt to standardize musical pitch standardization, see 143–148.

13 Sibille, *Harmony Must Dominate the World*, 61–82.

14 Union internationale de radiodiffusion. Letter from Charles Le Maistre to Edward Pawley, 2 May 1939. "Normal Pitch Standardisation 1937–1940 (Foreign general)." BBC, E2/437/1.

15 Since 1865, the International Telegraph Union regulated wireless communications. But this arrival of radio seemed to promise a new era of unrestricted communication that neither the International Telegraph Union nor the League of Nations wanted to restrain. As a result, European broadcasters took the initiative in creating the IBU, producing a non-governmental organization that might be capable of addressing this challenge. Suzanne Lommers, *Europe—On Air, Interwar Projects for Radio Broadcasting* (Amsterdam: Amsterdam University Press, 2012), chapter 1.

16 Lommers, *Europe—On Air*, 68.

17 Lommers, *Europe—On Air*, 20.

18 "les graves inconvénients qui résultent [des variations du *la*] tant pour l'échange des orchestres que pour les relais internationaux et pour toute la technique de la musique enregistrée." Union internationale de radiodiffusion. Document n° 447. Commission technique. Brussels, 15 Nov. 1937. Raymond Braillard, Exposé préliminaire. "Normal Pitch Standardisation 1937–1940 (Foreign general)." BBC, E2/437/1.

19 See Lommers, *Europe—On Air*, Chapter 6, "Broadcasting a Musical Culture," 235–287.

20 I borrow this phrase from Wittje, *The Age of Electroacoustics.*

21 On the role of engineers in interwar international relations and the process of European integration, see: Vincent Lagendijk, *Electrifying Europe: The Power of Europe in the Construction of Electricity Networks* (Amsterdam: Amsterdam University Press, 2008); by the same, "'To Consolidate Peace'? The International Electro-Technical Community and the Grid for a United States of Europe, 1929–1937," *Journal of Contemporary History* 47, no. 2 (March 2012): 402–426; Lommers, *Europe—On Air*; and Léonard Laborie, *L'Europe mise en réseaux: la France et la coopération internationale dans les postes et les télécommunications (années 1850-années 1950)* (Brussels: Peter Lang, 2011).

22 See especially Sibille, *Harmony Must Dominate the World.*

23 Lommers, *Europe—On Air*, 250ff.

24 Lommers, *Europe—On Air.*

25 Lommers, *Europe—On Air*, 247ff.

26 Notes of a meeting of ISA sub-committee no $43/3^{B}$, 11 and 12 May 1939.

27 Notes of a meeting of ISA sub-committee no $43/3^{B}$, 11 and 12 May 1939.

28 "International Standard Musical Pitch," *Journal of the Royal Society of Arts* 98 (1949): 76.

29 Notes of a meeting of ISA sub-committee no $43/3^{B}$, 11 and 12 May 1939.

30 Wittje, *The Age of Electroacoustics.*

31 Thompson, *The Soundscape of Modernity*, 105.

32 It was later renamed the Dutch Society for Acoustics. See Bijsterveld, *Mechanical Sounds*, 112; and, by the same, "A Sound Foundation: The Early Years of the Dutch Society for Acoustics," *Sound&Science: Digital Histories*, https://acoustics.mpiwg-berlin.mpg.de/contributor-essays/sound-foundation-early-years-dutch-society-acoustics.

33 Affaires traitées par le cabinet. 1925–1940. Note from the chief of staff of the Minister of Posts to the Minister of Posts. AN/19860208/47.

34 Thompson, *The Soundscape of Modernity*, 109.

35 Thompson, *The Soundscape of Modernity*, 104–105.

36 Thompson, *The Soundscape of Modernity*; and "A Sound Foundation."

37 Bijsterveld, *Mechanical Sounds*; and "A Sound Foundation."

38 Bijsterveld, *Mechanical Sounds*; and "A Sound Foundation."

39 Thompson, *The Soundscape of Modernity*, 155–166; Bijsterveld, *Mechanical Sounds*; Mills, "Deafening"; and by the same, "Decibel."

40 Bundesarchiv (hereafter "BArch") 4901/2741. The papers of another of these interwar acousticians, Vern Oliver Knudsen, cited below, similarly point to the international character of the careers of these technicians between the wars.

41 The first issue of the *Journal of the Acoustical Society of America* appeared in 1929. Likewise, between 1936 and 1942, Martin Grützmacher and Erwin Meyer edited the *Akustische Zeitschrift* on behalf of the German Science Foundation (DFG),

with the support of the Ministry of Posts and the Physikalisch-Technische-Reichsanstalt.

42 Craig N. Murphy, *International Organization and Industrial Change: Global Governance since 1850* (Oxford: Oxford University Press, 1994), 92–93, 196–197.

43 Craig N. Murphy, *International Organization and Industrial Change*, 56.

44 Yates and Murphy, *Engineering Rules*, 13.

45 Thompson, *The Soundscape of Modernity*, 355; and Rexmond C. Cochrane, *Measures for Progress: A History of the National Bureau of Standards* (Washington: U.S. Dept. of Commerce, 1966), 263 and Appendix G.

46 Archival material about the Bureau's work on acoustics can be found at the National Archives. RG 167. Records of the National Institute of Standards and Technology / National Bureau of Standards (Department of Commerce). General Correspondence, Boxes 13 (1918), 51 (1923), 94 (1924), 129 (1925), 168 (1926), 203A (1927), 221A (1928), 257 (1929), 296 (1930), 332 (1931), and so on.

47 Sub-Committee 26.

48 The committee was chaired by the acoustician G. W. G. Kaye and comprised the physicist J. H. Awbery from the National Physical Laboratory, the sound engineer Francis William Alexander, member of the research department at the BBC, the wind instrument maker Arthur J. Blaikley from the firm Boosey & Hawkes, the acoustician L. S. Lloyd, the piano and organ builder H. Willis, representing the Organ Manufacturers' Association, as well as the piano maker Evelyn Broadwood, speaking on behalf of his own company, but also a member of the Piano Manufacturers' Association. In order to secure a large adhesion to its decision, the BSI pitch committee organized a broad national consultation by publishing a letter in several periodicals "so that musical people with strong ideas on the subject could make themselves heard." Internal Circulating Memo from Alexander to the Overseas Engineering and Information Department, 6 June 1938. Broadwood papers, 2185/JB/71/4a.

49 Lommers, *Europe—On Air*, chapter 3, "Europe in the Making," 73–136.

50 Union internationale de radiodiffusion. Document n° 447. Commission technique. Brussels, 15 Nov. 1937. BBC, E2/437/1.

51 "groupe d'étude mixte." Union internationale de radiodiffusion. Document n° 447. Commission technique. Brussels, 15 Nov. 1937. Raymond Braillard, Exposé préliminaire. BBC, E2/437/1.

52 Union internationale de radiodiffusion. Letter [and translation] from von Hans-Joachim von Braunmühl to Hayes, 27 April 1938. Von Braunmühl was a central figure of the German radio. BBC, E2/437/1.

53 See below.

54 BBC. Alexander, Francis William. Personal File. Alexander joined the BBC in 1933 as an engineer in the Research Department. In April 1939, he became the head of the Programme Engineering Department. In 1938, his supervisor at the Research Department supported the choice of Alexander as ambassador of the BBC in the BSI pitch committee, arguing that "he ha[d] the great advantage of combined musical and technical knowledge." Memo from H. L. Kirke, 26 May 1938. BBC, E2/437/1.

55 "il avait équipé les studios avec une distribution centrale de la note 440 diffusée par haut-parleur et qu'il avait, en outre, réalisé des appareils portatifs." Letter from Braillard to Hayes, Brussels, 2 March 1940. BBC, E2/437/1.

56 "Au fond cette conférence était peut-être difficile mais ce fut quand même un 'bon temps' passé parmi de bons amis!" Letter from Braillard to Hayes, 29 April 1939. BBC, E2/437/1. This conference marked a very tense point in the history of the IBU. See Lommers, *Europe—On Air*, 150 and following.

57 "Reference was made by the Director to the close co-operation which the B.S.I. maintained with the standardising organisation in the United States, and said that it had been found that Anglo-American co-operation had always proved to have great weight in international discussions. He suggested that contact with the American Standards Association, with a view to finding out what the general practice of the United States was, so that, if possible, Great Britain and America could come to some agreement before any international discussion took place." BSI 2M (B) 2323. Sub-Committee B/26/7. Concert Pitch. Technical Committee B/26, Acoustics. Unconfirmed Minutes of 1st meeting held at the Offices of the Institution . . . 9 June 1938. Broadwood papers, 2185/JB/71/4a.

58 "Dr. Alexander said that, in order to obtain agreement with America, he felt that the B.B.C. would be prepared to consider the adoption of a slightly different pitch"; "it was agreed that the Director should get into touch with the American Standards Association to ascertain definitely what the American practice was." BSI 2M (B) 2323. Broadwood papers, 2185/JB/71/4a.

59 BSI. 2M (B) 2511. Sub-Committee B/26/7. Concert Pitch. Technical Committee B/26, Acoustics. Unconfirmed minutes of 2nd meeting held at the Offices of the Institution . . . 17 October 1938. Broadwood papers, 2185/JB/71/4a.

60 The meeting between the BSI's pitch committee and Knudsen also provided an opportunity for an exchange of knowledge. Knudsen returned to the United States with a lot of material on European negotiations. In turn, Knudsen shared American views on the definition of pitch with regard to temperature and gave details over the manner in which the Bureau of Standards broadcasted pitch by radio. Knudsen papers; Circulating Memo from Alexander to the Overseas and Engineering Information Department "For the Attention of Mr. Pawley," 2 Dec. 1938. BBC, E2/437/1.

61 Born and raised in Heidelberg, Lottermoser attended school in Danzig and Münster, where he obtained a Doktorat in physics with a work entitled *Über die Dielektrizitätskonstante von Flüssigkeitsgemischen* in 1924.

62 Letter from the German Secretariat to the members of Sub-Committee 3. Electroacoustics, Musical Acoustics of ISA 43. Acoustics, and Report in the Measurement of the Concert Pitch a' in Musical Broadcasts, Submitted by Werner Lottermoser, Phys.-Techn. Reichsanstalt, Berlin. Broadwood papers, 2185/JB/71/4a.

63 David Cahan, *Meister der Messung. Die Physikalisch-Technische Reichsanstalt im Deutschen Kaiserreich* (Bremerhaven: Wirtschaftsverl. NW, Verl. für Neue Wiss., 2011), 183–184; and Jackson, *Harmonious Triads*, 226–227.

64 See Ulrich Kern, *Forschung und Präzisionsmessung. Die Physikalisch-Technische Reichsanstalt zwischen 1918 und 1948* (Weinheim and New York: VCH Verlagsgesellschaft, 1994); and Dieter Hoffmann, "Die Physikalisch-Technische Reichsanstalt im Dritten Reich," *PTB Mitteilungen* 122, no. 2 (2012): 30–31.

65 BArch 1519/825; and BArch 1519/827: "Akustische Minenzündung.- Schriftwechsel der PTR mit dem Institut für Schwingungsforschung und dem Reichsministerium für Lufthaft."

66 Kern, *Forschung und Präzisionsmessung*; on sound and propaganda under the Third Reich, see Carolyn Birdsall, *Nazi Soundscapes: Sound, Technology, and Urban Space in Germany, 1933–1945* (Amsterdam: Amsterdam University Press, 2012).

67 Grützmacher's first published article engaged with the theories of Helmholtz and Carl Stumpf, offering a new method for sound analysis. "Eine neue Methode der Klanganalyse," *Zeitschrift für technische Physik* 8, no. 11 (Nov. 1927): 506–509.

68 "Eine neue Methode der Klanganalyse. Elektrische Nachrichtentechnik (ENT)," *Zeitschrift für technische Physik* 4, no. 12 (Dec. 1927): 533–545.

69 Martin Grützmacher and Werner Lottermoser, "Uber ein Verfahren zur trägheitsfreien Aufzeichnung von Melodiekurven," *Akustische Zeitschrift* 2 (1937): 242; and "Die Verwendung des Tonhöhenschreibers bei mathematischen, phonetischen und musikalischen Aufgaben," *Akustische Zeitschrift* 2 (1938): 183. Archival material about this study can be found in BArch R 73/11361.

70 The study was entitled "Statistische Untersuchungen über charakteristische Unterschiedsmerkmale von Flügeln, Orgel und Blasinstrumenten." The results of these experiments were published in the following papers: "Über die Stimmungen von Flügeln," *Physikalische Zeitschrift* 36, no. 24 (11 Oct. 1935): 903–912; and "Neuere Untersuchungen an Flügeln," *Akustische Zeitschrift* 1 (1936): 49–58.

71 Werner Lottermoser, *Orgeln, Kirchen, und Akustik* (Frankfurt am Main: Bochinsky, 1983).

72 "Dr. Werner Lottermoser ist als ausübender Musiker besonders an der Bearbeitung musikalisch-akustischer Probleme interessiert, und hat zusammen mit dem Unterzeichneten das Stimmungsproblem von Musikinstrumenten bearbeitet." Letter from Grützmacher to the Deutsche Forschungsgesellschaft, 3 Jan. 1936. BArch R73/12785.

73 "Es werden beträchtliche, doch vorwiegend unsystematische Abweichungen von der mathematisch temperierten Stimmung bis zu einigen Hertz festgestellt, die aber vom Ohr keineswegs als Fehler, sondern als Bereicherung des Klanges gehört werden." "Significant, but mostly unsystematic deviations from the mathematically tempered tuning of up to a few Hertz are found, which are by no means heard by the ear as a mistake, but as an enrichment of the sound." Grützmacher and Lottermoser, "Über die Stimmungen von Flügeln."

74 "Der Einfluß des Materials von Orgel-Metallpfeifen auf ihre Tongebung." *Akustische Zeitschrift* 2 (May 1937): 134.

75 Further research remains to be done to determine Lottermoser's level of adhesion to Nazi politics, but as a file held at the Bundesarchiv reveals, he has been a member of Nazi Party since 1 May 1932. BArch, R/9361/I. Parteistatistische Erhebung 1939 18.6.09. (*Mitgliedsnummer* [membership number]: 1135698).

76 "During the measurements, or during the tuning of the instruments, was used for bringing the standard tone." Letter from the German Secretariat to the members of Sub-Committee 3. This filtering activity is reminiscent of what the music theorist Heinrich Schenker called "primary tones."

77 Robert Brain, *The Pulse of Modernism: Physiological Aesthetics in Fin-de-siècle Europe* (Seattle: University of Washington Press, 2015); and Soraja de Chadarevian, "Graphical Method and Discipline: Self-Recording Instruments in Nineteenth-Century Physiology," *Studies in the History and Philosophy of Science* 24, no 2 (1993): 267–291.

78 Brain, *The Pulse of Modernism*, 28.

79 Letter from the German Secretariat to the members of Sub-Committee 3.

80 According to Dumont, Gustave Lyon, an employee of Pleyel and the acoustician responsible for the room's acoustics, led a similar experiment. Association française de normalisation. Sept. 1938. Doc—Ac—501. Acoustique Musicale. Normalisation du Diapason. Note de Monsieur A. Dumont. BArch 4901/2741.

81 BBC, E2/437/1. On Tatsfield Receiving Station, see Edward Pawley, *BBC Engineering*, 194 and following.

82 BBC. Union Internationale de Radiodiffusion. Document n° 482. Mesures du la normal, effectivement utilisé par les orchestres de la RRG, et quelques points de vue pour la fixation nouvelle. 16 June 1838. BBC, E2/437/1.

83 Appendix to ISA Circular 43/3ᴮ 01 (24 April 1939). Commission internationale de radiodiffusion. Bruxelles, le 22 décembre 38. BArch 4901/2741.

84 "avec une précision de 0,2 périodes/sec." Appendix to ISA Circular 43/3ᴮ 01 (24 April 1939). Commission internationale de radiodiffusion. Bruxelles, le 22 décembre 38. BArch 4901/2741.

85 BSI CF (B) 2555. French Acoustical Committee. Contribution to the study of the normal "A." Technical Note on measurements of the level of the normal "A" of broadcasting orchestras. by M. A. Labrousse—Engineer of the French Post Office, 10 May 1939. Broadwood papers, 2185/JB/71/4a.

86 "The currents originating either from a modulation centre, through a telephone line, or from a broadcasting receiver, were made to pass through the pass-band filter and, after suitable amplification, were recorded on an oscillographic tape at the same time as the standard frequency of 1000 c/s (supplied by the Laboratoire National de Radioélectricité)." BSI CF (B) 2555. French Acoustical Committee. Contribution to the study of the normal "A." Technical Note. Broadwood papers, 2185/JB/71/4a.

87 BSI CF (B) 2555. French Acoustical Committee. Contribution to the study of the normal "A." Technical Note. Broadwood papers, 2185/JB/71/4a.

88 Appendix to ISA Circular 43/3ᴮ 01 (24 April 1939). Commission internationale de radiodiffusion. Bruxelles, le 22 décembre 38. BArch 4901/2741. The Bach concerto is probably rather the C-minor concerto for violin and oboe BWV 1060R.

89 Elena Aronova, Christine von Oertzen, and David Sepkoski, "Introduction: Historicizing Big Data," *Osiris* 32 (2017): 1–17.

90 BSI CF (B) 2555. French Acoustical Committee. Contribution to the study of the normal "A." Technical Note. Broadwood papers, 2185/JB/71/4a.

91 Broadwood papers, 2185/49/31 and 32. The program of this itinerant outdoor concert was: William Byrd (1543–1623), "This sweet and merry month of May"; John Wilbye, "Sweet honey-sucking bees"; Thomas Tomkins, "See, see, the Shepherd's Queen"; "Brigg Fair," folk song arranged by Percy Grainger; "Wassail" and "The

Turtle Dove," two folk songs arranged by Vaughan Williams; Anon., "Cuckow"; Orlande de Lassus, "Thou knowest fairest maiden" and "Un jour vis un foulon"; and Orlando Gibbons, "What is our life?" It was followed by an indoor performance of popular songs and one early nineteenth-century song. "Reception at Lyne. Capel. Surrey to meet the Delegates of the International Committee on the Standardisation of Concert Pitch (International Standards Association) invited to England by the British Standards Institution. Saturday, May 13th, 1939." Broadwood papers, 2185/JB/71/4a.

92 BBC, E2/437/1. Letter from Ethel R. Wiggins to Pawley, London, 1 May 1940. The documents produced by BSI made the diffusion of standards adopted by the organization a priority. See for instance: "NOTE. Co-operation within the Empire. In accordance with the agreements established at successive Imperial Conferences, information in regard to all new work started is circulated to the National Standardising organisations in Australia, New Zealand, South Africa and Canada and a copy of these Minutes will be forwarded 7 days after their issue to those bodies for information. Please therefore submit at once in writing any correction you think necessary to the Minutes." BSI 2M (B) 2323, 16 June 1938. Note: Cooperation within the Empire. Broadwood papers, 2185/JB/71/4a. The report required members of the committee to pay attention to this question: BSI document about the application of BSI recommendation to Empire. When sending out this form to the members of the pitch committee, the BSI acoustician Weston wrote that "[their] attention is called to the note at the top of these Minutes." Circular letter from Weston to the members of the subcommittee concert pitch. 13 July 1938. Broadwood papers, 2185/JB/71/4a.

93 ISA Committee 43/3[B] submits the following recommendation to the National Standardising Bodies affiliated to the ISA also to the C.C.I.F., the I.E.C. and the U.I.R.: 1. That the international standard of concert pitch be based on a frequency of 440 cycles per second for the note A in the treble clef. 2. That this value be maintained within the closest limits possible by soloists, orchestras, choirs etc. throughout all musical performances, and also in recorded music. 3. That with a view to reducing the necessary tolerances to acceptable values, a set of technical recommendations be adopted, preferably on the basis of international cooperation.

94 "[D]es postes émetteurs radiophoniques." Document n° 580 U.I.R. Contribution pour une réponse aux questions mises à l'étude par la réunion du Comité ISA 43/3[B] à Londres concernant l'observation exacte de la fréquence d'accord de 440 c/s pour le "la" normal. (Annexe aux résolutions de Londres). BBC, E2/437/1.

95 Radio engineers were not the only technical personnel concerned with the production of the signal of standard pitch. Telephone ringtones were also considered to be appropriate vehicles of the new standard. This question was raised in the proceedings of the 1939 conference. Engineers from IBU, however, dismissed the idea of diffusion of "the tuning note through the telephone in a way similar to the one used for the distribution of exact time could be studied; however the relatively weak intensity of telephone signals seems to be a disadvantage of this procedure." "La diffusion de la note d'accord par le téléphone d'une manière analogue à celle utilisée pour la distribution de l'heure exacte pourrait être également étudiée; toutefois l'intensité sonore relativement faible des signaux téléphoniques semble être un désavantage de cette procédure." Document n° 580 U.I.R. Contribution pour une réponse aux questions mises à l'étude par la réunion du Comité ISA 43/3[B] à Londres concernant l'observation exacte de la fréquence d'accord de

440 c/s pour le "la" normal. (Annexe aux résolutions de Londres), 5 June 1939. BBC, E2/437/1.

96 Memo from Hayes, 22 Dec. 1937. BBC, E2/437/1.

97 "I am delighted to hear Mr. Hayes's suggestion that a standard of 'A' should be available for us in the loudspeakers of all studios. You will remember that Mr. Stanford Robinson asked for this some years ago, and I thought it a very good idea, but at the time I think you found it impracticable. It will be of the greatest value to us all if it can be done." Memo from Boult, 24 Dec. 1937. BBC, E2/437/1.

98 "J'avais demandé . . . lorsqu'on nous a construit notre horloge spéciale pour les signaux horaires de la radiodiffusion suisse que ces signaux soient donnés sur la fréquence de 440 Hertz, soit sur le 'la' du diapason, de telle façon qu'ils puissent être utiles non seulement à ceux qui veulent régler leur montre mais également aux musiciens et aux accordeurs de pianos. . . . Cette idée [est] probablement plus importante qu'elle n'en a l'air puisqu'il s'agit pour la radiodiffusion de mettre tout le monde d'accord en donnant le *la* au concert européen." Letter from Maurice Rambert to Balthasar van der Pol, Geneva, 4 April 1939. Document n° 574 U.I.R. Diffusion du "la normal," 17 May 1939. BBC, R 54/21/1.

99 Notes of a meeting of ISA subcommittee no $43/3^{B}$, concert pitch, 11 and 12 May 1939.

100 See below, chapter 5.

101 BSI. 2M (B) 2511. Unconfirmed Minutes of second meeting held at the Offices of the Institution . . . 17 Oct. 1938. Broadwood papers, 2185/JB/71/4a.

102 "In the course of the discussion Professor Knudsen informed the Committee that the standard American pitch of 440 c/s is broadcast every ten minutes by Broadcasting Station of the American National Bureau of Standards." Internal Circulating Memo from Alexander to Overseas Engineering and Information Department, "For the Attention of Mr. Pawley," 2 Dec. 1938, BBC, E2/437/1.

103 Internal Circulating Memo from Alexander to Overseas Engineering and Information Department. BBC, E2/437/1.

104 See below, chapter 5.

105 "Such a piece of apparatus would also be useful for keeping our organs to the right tuning fork and checking upon tuning forks. This standard pitch tuning fork should also be used for lining-up the Empire transmissions, due to the widely varying temperatures experienced through the Empire and the necessity of maintaining this absolute pitch, irrespective of the temperature." BBC, E2/437/1.

106 Pawley, *BBC Engineering*, 125. On the adoption of *A* 439 in Britain, see above, chapter 2.

107 The Assistant Controller of Engineering suggested that the accuracy described by Alexander of 0.1 cycles per second would be hard to reach. Memo from the Assistant Controller of Engineering, 21 Jan. 1938. BBC, E2/437/1. This issue was finally solved by the use of a crystal-controlled oscillator. F. W. Alexander, "Standard Musical Pitch," *BBC Quarterly* 6, no. 1 (Spring 1951): 62.

108 Memo from Hayes, 12 June 1939. BBC, E2/437/1.

109 Memo from Alexander to Hayes and Pawley, 13 June 1939. BBC, E2/437/1.

110 Memo from Alexander to Hayes and Pawley, 13 June 1939.

111 Document n° 574 U.I.R. Diffusion du "la normal," 17 May 1939. Letter from van der Pol to Rambert, 13 April 1939, BBC, R 54/21/1.

112 Letter from Pawley to Le Maistre, 13 Sept. 1939, BBC, R 54/21/1.

113 Document n° 574 U.I.R. Diffusion du "la normal," 17 May 1939. Letter from van der Pol to Rambert, 13 April 1939, BBC, R 54/21/1.

114 Letter from Hayes to unidentified receiver, 23 Feb. 1940. Hayes does go on to qualify this strong sentiment with "I do, however, agree that it should take its correct place in the matter of priority, i.e., fairly low down." BBC, E2/437/1.

115 https://www.vice.com/en/article/xywy74/the-fringe-audiophiles-who-want-to-topple-standard-tuning.

116 On the genealogy of conspiracy theories surrounding *A* 440, see my article "Tuning Forks as Time Travel Machines" and below, chapter 5.

CHAPTER FIVE

Letter from Lloyd S. Lloyd to Edward Pawley, 16 Feb. 1947. BBC, R54/21/2, "Technical General. Personal: BSI. Standard Pitch. 1947."

1 JoAnn Yates and Craig N. Murphy, *Engineering Rules: Global Standard Setting since 1880* (Baltimore: Johns Hopkins University Press, 2019); and Craig N. Murphy and JoAnn Yates, *The International Organization for Standardization (ISO): Global Governance Through Voluntary Consensus* (Oxford: Routledge, 2008).

2 Letter from Ethel Wiggins to Pawley, 11 Dec. 1945. BBC, R54/21/1, "Technical General. Personal: BSI. Standard Pitch. 1937–1946."

3 Letter from Wiggins to Pawley, 11 Dec. 1945. BBC, R54/21/1.

4 This included the broadcast of an *A* 440 signal as part of the BBC's daily programs, and the installation of electronic tuning devices supplying the new tone in all studios. See above, chapter 4.

5 Letter from Pawley to Wiggins, 19 Dec. 1945. BBC, R54/21/1. On receiving Wiggins's letter, Pawley forwarded it to the head of Overseas Engineering and International Development at the BBC, Hayes, with the following note: "Can you help please. It would be a pity if the agreement were allowed to lapse, after all the work that was done." Letter from Wiggins to Pawley, 11 Dec. 1945. BBC, R54/21/1.

6 Memo from Pawley, 21 Dec. 1945. BBC, R54/21/1.

7 In August 1951, claiming that there were no "high grade" tuning forks in Australia, and eager to emulate the BBC's initiative to equip their piano tuners with standard tuning forks, the ABC ordered two tuning forks and two tuning gongs from Boosey & Hawkes in London. Not only did it take several months for the two tuning forks to arrive, in mid-1952, by sea mail, but these devices were found to be inaccurate and, thus, sent back to London by freight. The ABC only received a new order of tuning forks on 15 August 1952, while the two gongs were delivered to the wrong address and thus had to be ordered again. Purtell, *Tuning the Antipodes*, 156.

8 Purtell, *Tuning the Antipodes*, 157.

9 BSI. Technical Committee ACM/7—Musical Pitch. Further Reply from Standardisation Department, Mexico, Concerning Implementation of the Decisions

of the 1939 Conference on Concert Pitch. BBC, R54/21/3, "Technical General. Personal: BSI. Standard Pitch. 1948–1950."

10 "Melba Had a Say in Our 'A'," *Argus* (6 July 1950): 5. Quoted in Purtell, *Tuning the Antipodes*, 195. On Melba and the diapason normal, see above, chapter 2.

11 ISO Technical Committee 43 Acoustics. German Proposal Concerning the standardization of Musical Pitch, 25 October 1953. BBC, E2/437/2.

12 Letter from Wiggins to Pawley, 4 Sept. 1946, BBC, R54/21/3. The article that Lloyd referred to is "Measurements of Orchestral Pitch" (Bell Telephone Systems Monograph B-1282. It is a reprint of O. J. Murphy, "Measurements of Orchestral Pitch," *Journal of the Acoustical Society of America* [12 Jan. 1941]: 395–398). Letter from Pawley to Wiggins, 28 Oct. 1946. BBC, R54/21/3.

13 Maida Vale's Studio 1.

14 BSI. 2M (B) 9533. Unconfirmed Minutes of a meeting held at the Offices of the Institution (on 2 July 1946). BBC, R54/21/1.

15 BSI. 3M (B) 748. Unconfirmed Minutes of meeting held at the Offices of the Institution, (on 26 March 1947). BBC, R54/21/2.

16 Memo from Pawley, 27 March 1947. BBC, R54/21/2.

17 Memo from Hotine to Pawley, 12 July 1946. BBC, R54/21/1.

18 Memo from Pawley to Hotine, 12 July 1946. BBC, R54/21/1.

19 The social implications of concert pitch *A* 440's broadcast on the radio were also manifest through the care that Pawley took in making sure that this tone could not create the opposite effect and introduce some confusion among listeners. As much as the engineer perceived the benefits of broadcasting concert pitch to tune audiences' ears, he feared the detrimental effect that the broadcast of a slightly out-of-tune signal may have. This is why he recommended that in case the 440 note could not be properly emitted, they should switch to an entirely different tone, emphasizing that "it is important that if the accurate 440 c/s note is not available a note of a quite different frequency (e.g. 1000 c/s) should be used so as to eliminate any possible confusion." Memo from Pawley to H.O.E.I.D., 27 March 1947. BBC, R54/21/2.

20 BSI. Unconfirmed Minutes of a meeting held at the Offices of the Institution (on 2 July 1946). BBC, R54/21/1.

21 Blaikley, Lloyd, and Pawley.

22 "Can we have an 'A'?." BBC, R54/21/2.

23 Pawley immediately suggested that the chairman of the BSI pitch committee, Lloyd S. Lloyd, "be asked to give such a talk (with suitable illustrations)," as he confessed that he did "not know whether he would be a good broadcaster." Memo from Pawley, 27 Feb. 1947. BBC, R54/21/2. Pawley later suggested that "the talk might be given in the form of an interview between a good broadcaster like Malcolm Sargent and the Chairman of the B.S.I. Committee, Mr. Ll. S. Lloyd, C.B., and should be accompanied by illustrations." Memo from Pawley to A.C.E., 4 March 1947, BBC, R54/21/2.

24 "Can we have an A?," 17–18. BBC, R54/21/2.

25 Many members of the committee missed the show, so Wiggins organized a replay at the BBC for the members of the body. Letter from Wiggins to the members of

the BSI pitch committee, 10 June 1947. BBC, R54/21/2. On the organization of this event, see Wiggins's subsequent exchange with Pawley on this matter in the same file.

26 Letter from Pawley to Banister, 29 April 1949. BBC, R54/21/3; *A Seat at the Proms: For Lovers of the Orchestra* (London: Evans Brothers Limited, 1948).

27 In his report of the BSI committee meeting of 7 Oct. 1949, for example, Stanley, the assistant technical director of the BBC, wrote to Pawley that the committee would "appreciate if you could publicize the pitch and state what it is for in the appropriate broadcasting literature" and "agreed that other countries should be informed that the broadcasting to the Standard pitch in the United Kingdom has had beneficial results, and that they should be advised to broadcast the Standard pitch in their countries." Letter from Stanley to Pawley, 30 Dec. 1949. BBC, R54/21/3, "Technical General. Personal: BSI. Standard Pitch. 1948–1950." Pawley suggested that "the wording might be as follows": "The tuning note broadcast at the start of the Third Programme each evening is correctly maintained at the international standard of pitch. It has a frequency of 440 cycles per second, and corresponds to the note a in the treble stave. It is hoped that it will be useful to performers and makers of musical instruments by enabling them to turn their instruments to the standard pitch." Memo from Pawley, 5 Jan. 1950. BBC, R54/21/3.

28 Ottavio Tiby and Alfonso Barone, *Note e rilievi sulla frequenza del "La$_3$"* (Rome: Ministerio della cultura popolare, 1941). On Ottavio Tiby, see Guglielmo Barblan and Renate Albrecht, "Ottavio Tiby (19. 5. 1891 bis 4. 12. 1955)," *Die Musikforschung* 9, no. 3 (1956): 296–297. He was born in Palermo, where he graduated in composition in 1921. An expert of Sicilian popular music, Tiby also published on Byzantine music, theory, and history (1938), and music in Greece and Rome (1942). From 1940 to 1944 he was a teacher of acoustics and organology at the Santa Cecilia Conservatory in Rome.

29 O. J. Murphy, "Measurements of Orchestral Pitch." As mentioned above, AFNOR's director, Duval, referenced this study during one of the meetings of the body's technical committee on pitch.

30 BSI. CK (ACM) 6266, Technical Committee ACM/7—Concert Pitch. Replies from Standards Bodies Concerning Implementation of the Decision of the 1939 Conference on Concert Pitch. 3 Oct. 1949. BBC, R54/21/3.

31 The comparatively good performance of brass bands that Murphy's study revealed was at the center of this article: "Brass Bands Better Than Orchestras—At least in keeping to the American Standard pitch, Bell Laboratories' engineer finds," *Industrial Standardization and Commercial Standards Monthly* 12 (July 1941): 186.

32 They also quoted the answer of an orchestra leader "whose orchestra uses the 442 Hz pitch" but "indicated his good intentions" by claiming: "'We try to get the musical pitch down to 440.'" ISO Technical Committee 43 Acoustics. German Proposal Concerning the Standardization of Musical Pitch. BBC, E2/437/2.

33 "before the performance a certain number of standard 'la's were chosen by the orchestra from the composition or compositions that were to be performed; and by means of the radio apparatus the performance by the section of the orchestra was followed. Approaching one of the standard 'la's marked, the sensitive sheet of the oscillograph was set running; at the precise moment at which the recording of

the frequency occurred, by means of a touch a signal was transmitted to a second oscillograph equipment, so as to make an identification signal corresponding with the standard 'la.'"

34 BSI CH (B) 8630. Extracts from a paper entitled "Notes and data on the frequency for standard 'la,'" 20 Jan. 1947, BBC, R54/21/2.

35 BSI CK (ACM) 9996. Technical Committee ACM/7—Musical Pitch. Further Reply from Standardisation Department, Mexico, Concerning Implementation of the Decisions of the 1939 Conference on Concert Pitch. 9 March 1950. BBC, R54/21/3. The report was the work of a committee "on the adoption of note LA3, with 440 vibrations per second," at the National School of Music at the Autonomous University in Mexico (the Escuela Nacional de Musica, of the Universidad Autonoma de Mexico) which the Secretariat of Public Education had appointed through the "Dirección General de Normas," a member of ISO.

36 Extracts from a paper entitled "Notes and data on the frequency for standard 'la.'"

37 Carl Seashore, *Psychology of Music* (New York and London: McGraw-Hill Company, 1938).

38 Extracts from a paper entitled "Notes and data on the frequency for standard 'la.'"

39 BSI CH (B) 8630. Extracts from a paper entitled "Notes and data on the frequency for standard 'la,'" 20 Jan. 1947, BBC, R54/21/2.

40 BSI 3M (B) 748. Unconfirmed Minutes of meeting held at the Offices of the Institution [on 26 March 1947], 18 April 1947. BBC, R54/21/2.

41 "Le PRÉSIDENT souligne la nécessité d'une marque de conformité, sans laquelle les spécifications perdraient de leur intérêt. M. ACOULON fait remarquer qu'elle ne peut concerner que les instruments à clés: flûte, hautbois, clarinette, etc. . . . Sur la demande du Président il indique les caractéristiques qui pourraient être l'objet d'un contrôle: la justesse (exactitude de tous les écarts) et la précision de l'accord. Celle-ci peut être obtenue, pour les instruments à clé, à +/- 0,5 périodes à 20°." AFNOR. Procès-verbal. Deuxième réunion du comité restreint de l'Acoustique, May 1950. AN, AJ/37/486, "Fixation du diapason: avis, rapports, correspondance. 1933–1936 et 1950–1954."

42 3M (B) 3777. Sub-Committee B/26/7—Concert Pitch of Technical Committee B/26—Acoustics. Unconfirmed Minutes of meeting, 31 Jan. 1949. BBC, R54/21/3.

43 "les fabricants signent leurs instruments et donc en prennent la responsabilité . . . un instrument tel qu'une clarinette peut varier d'un quart de ton lorsque l'exécutant change." AFNOR. Procès-verbal. Deuxième réunion du comité restreint de l'Acoustique, May 1950. AN, AJ/37/486.

44 Robert Dussaut (1896–1969), Rémy Campos, "L'Enseignement de la composition au Conservatoire de Paris," in Laurent Feneyrou and Alain Poirier (eds.), *De la Libération au Domaine musical: Dix ans de musique en France (1944–1954)* (Paris: Vrin, 2018), 146.

45 In 1954, for example, he claimed that he had alerted musical circles about pitch's escalation since 1945, while in 1950, he had alleged that he had been working on the question for two years. Letter from Robert Dussaut to Jacques Jaujard, general director of Arts and Letters, 13 June 1954. AN, AJ/37/486; "Discussion sur le diapason," 9 April 1950. AN, AJ/37/486. "Fixation du diapason: avis, rapports,

correspondance. 1933–1936 et 1950–1954." The first record I have found of Dussaut's public engagement with the question of pitch standardization is this article: "Diapason trop haut!" Musique et Radio. *Revue technique et professionnelle de musique* 39, no. 444 (May 1948): 145–147.

46 After an initial letter from February 1949, the BSI wrote to ISO members in January 1950 to ask for information and announce their intention to call for an international conference. BSI, CK (ACM) 8682. Technical Committee ACM/7—Concert Pitch. BBC, R54/21/3. As mentioned above, the BSI first made contact with AFNOR and other national standard-setting bodies, following the publication of the *Herald Tribune* article, on 11 Dec. 1945.

47 Institut International du Son. Invitation to the first meeting dedicated to the discussion of the new French pitch, to Alice Viardot. AN, AJ/37/486.

48 Bibliothèque nationale de France. Fond Robert Dussaut. File 1.

49 "Il regrette que les musiciens n'aient pas été consultés en 1939 car ils se seraient opposés à l'adoption de la fréquence 440." AFNOR, Procès-verbal. Première réunion du Comité restreint de l'acoustique, April 1950. AN, AJ/37/486.

50 Letter from Dussaut to Delvincourt, 29 April 1950. AN, AJ/37/486.

51 "la signature, obtenue ensuite à Paris et par surprise, d'Henri Rabaud n'a pas été faite au nom du collectif de tous les musiciens français et . . . d'ailleurs, Henri Rabaud n'a pas été invité à assister aux délibérations du Congrès." Institut de France. Résolution adoptée par l'Académie des Beaux-Arts. Académie des sciences, "Commission du diapason." Archival evidence is lacking to support or invalidate Dussaut's claim. In the text Dussaut submitted to the Academy of Fine Arts (see below), he claimed that it was an instrument maker who collected Rabaud's signature. Communication à Messieurs les Membres de la Section de Musique de l'Académie des Beaux-Arts. Discussion sur le diapason, 9 April 1950. AN, AJ/37/486.

52 "pour le La 432 j'ai obtenu des centaines de signatures, et des milliers d'adhésions dans toute la France. J'avais ainsi démontré que les musiciens en général, désapprouvaient *le* La *440*." Letter from Dussaut to Delvincourt, 30 March 1951. AN, AJ/37/486 (Dussaut's emphasis).

53 AFNOR, Procès-verbal. Première réunion du Comité restreint de l'acoustique, April 1950. AN, AJ/37/486.

54 "La fréquence moyenne s'établit à 441, 3 périodes à seconde pour l'ensemble des concerts d'orchestre ou d'instruments solos." AFNOR, Procès-verbal. Première réunion du Comité restreint de l'acoustique, April 1950. AN, AJ/37/486. Duval probably refers to the following article: "Measurements of Orchestral Pitch."

55 "depuis 1939 tout l'outillage a été adapté à la fréquence 440 et qu'un nouveau changement entraînerait des dépenses considérables." AFNOR, Procès-verbal. Première réunion du Comité restreint de l'acoustique, April 1950. AN, AJ/37/486.

56 Letter from Dussaut to Delvincourt, 9 April 1950. AN, AJ/37/486.

57 Letter from Dussaut to Delvincourt, 29 April 1950. AN, AJ/37/486.

58 "il faut créer une ambiance de baisse"; "du temps de Bach par exemple les instruments n'étaient pas les mêmes qu'actuellement et les instrumentistes ne valaient pas ceux d'aujourd'hui. La montée du diapason a donc répondu à un certain

besoin." AFNOR, Procès-verbal. Deuxième réunion du Comité restreint de l'acoustique, May 1950. AN, AJ/37/486.

59 In his letter to Delvincourt, Dussaut insisted on AFNOR's "bad faith." According to him, Bernhart was guilty of committing "blackmail" by threatening to broadcast exclusively concerts at *A* 440. Dussaut also alleged that what he said during the last meeting was "deformed" in the *réunion*'s minutes. Finally, he pointed out the fact that the letter which Henri Busser had sent to AFNOR on 21 April 1950 to endorse Dussaut's proposed reform was not read. Letter from Dussaut to Delvincourt, 29 April 1950. AN, AJ/37/486.

60 "les musiciens sont guidés dictatorialement par des 'techniciens' (fabricants et ingénieurs de la Radiodiffusion) qui n'ont pas d'oreille musicale."

61 "pour adopter les degrés de l'échelle thermométrique, a-t-on demandé leur avis aux fabricants de thermomètres?" "Fragment d'un article de P.-J.Richard," Dussaut, Communication à Monsieur le Directeur général des Arts et des Lettres. La querelle du diapason, 28 Nov. 1951. AN, 20140260/16, Direction générale des Arts et des Lettres. 19341961, "Commission chargée de la hauteur du diapason (créée par arrêté du 6 août 1951): correspondance, rapport d'activité (1950–1952)."

62 Dussaut sent this through his former teacher and friend Henri Busser, who was a member of the institution.

63 "le Docteur Grutmacher et le Comité acoustique de la Radio de Berlin"; "En Septembre 1938 (au moment de Munich!)." Dussaut, Communication à Messieurs les Membres de la Section de Musique de l'Académie des Beaux-Arts. Discussion sur le diapason, 9 April 1950. AN, AJ/37/486.

64 "Ce sont donc les trompettes et saxophonistes nègres qui imposent à la vieille Europe un diapason grâce auquel nos chanteurs s'égosillent et les trésors de notre musique symphonique et théâtrale deviennent inentendables." René Dumesnil, "Musique," *Le Mercure de France*, no. 1042 (1 June 1950): 331. Dussaut's involvement in the publication of this article is all the more probable that fragments of this article (including the quoted passage) were copied in "Le diapason logique (la3=432). Attestation en faveur de la thèse des musiciens." AN, AJ/37/486. At the very least, this shows that Dussaut fully endorsed Dumesnil's rhetoric.

65 "Grâce à la contagion de leur musique populaire nègre, on peut dire que [la 440] est devenu la norme universelle." René Sudre, "Revue scientifique. Les incartades du diapason," *Revue des Deux Mondes* (15 Oct. 1950): 752.

66 "renoncer au franc, au mètre, à la langue française?" Dussaut, Communication à Monsieur le Directeur général des Arts et des Lettres.

67 Paul Bastide (1879–1962), a composer and conductor, was the head of music studies at the Opéra-Comique by the time Dussaut wrote this text. Dussaut, Communication à Messieurs les Membres de la Section de Musique de l'Académie des Beaux-Arts.

68 Dussaut, Communication à Messieurs les Membres de la Section de Musique de l'Académie des Beaux-Arts.

69 Busser again served as an intermediary between Dussaut and the institution: it was him who requested an audience from its members, and him again who introduced Dussaut's note along with the physician, biologist, botanist, and member of the scientific academy Joseph Magrou, on 19 June 1950. Letter from Henri

Busser to Louis de Broglie, perpetual secretary of the Academy of Sciences, 4 June 1950. Académie des sciences, "Commission du diapason." The text of this note was reproduced under the title "Proposition d'un nouveau son fixe: Sol_3=-fréquence 384. D'où La_3=432," in the *Comptes Rendus hebdomadaires des séances de l'Académie des sciences*, t. 230, 2150–2152.

70 See above, chapter 1.

71 On 22 June 1950, Busser forwarded the vow the Academy of Fine Arts had adopted the day before to the Academy of Sciences, asking whether the latter "could join us by expressing the same vow to Mister the Minister of Education." "se joindre à nous en exprimant le même vœu à Monsieur le Ministre de l'Éducation Nationale." Busser to [one of the perpetual secretaries of the Academy of Sciences], 22 June 1950. Académie des sciences, "Commission du diapason." The Academy adopted the text of the vow on 3 July 1950 and sent it to the minister of education the following day.

72 Directed by Paul Léon, president of the Union des Arts, the commission was composed of Albert Pérard, a member of the Academy of Sciences and of the Bureau des Longitudes, Delvincourt, Busser, two composers (including Dussaut), two music teachers, two conductors, two singers, two instrumentalists, two directors of musical associations, as well as three representatives of the musical industry.

73 "depuis sa création . . . [la commission] n'a absolument rien entrepris." Letter from Dussaut to Jaujard, 13 June 1954. AN, AJ/37/486. In a letter from October 1952, the secretaries of the Academy of Sciences wrote that since the fall 1951, "one has to admit that the situation has not made much progress." "Il faut bien reconnaître que la situation n'a pas beaucoup progressé." The perpetual secretaries of the Academy of Sciences to William Braid White, 19 Oct. 1952. Académie des sciences, "Commission du diapason." The composer was not able to attend the commission's first meeting of the commission, and its director then failed to organize another meeting. What is more, over the course of the following months, French musicians' alleged mobilization in favor of his reform appeared to have faded. See Dussaut, Discussion sur le choix le plus rationnel du la 3, April 1952. AN, AJ/37/486.

74 Dussaut to Delvincourt, 8 Oct. 1951. AN, AJ/37/486.

75 "les conclusions . . . du congrès de Londres n'auront aucune valeur aux yeux des musiciens en général et des musiciens français en particulier." Letter from Delvincourt to Dussaut, 13 Oct. 1953. AN, AJ/37/486.

76 Léon and Pérard.

77 Dussaut, Copie de la lettre de protestation qui avait été adressée le 16 octobre 1953, à l'AFNOR et à la British Standards Institution . . . par le Secrétariat d'État aux Beaux-Arts. AN, AJ/37/486.

78 "Les musiciens et physiciens français ne sauraient accepter la décision du prochain Congrès de Londres." Copy of the letter of protest sent to AFNOR and the BSI, 16 Oct. 1953. AN, AJ/37/486 (Dussaut's emphasis).

79 "l'intervention du Secrétariat d'État aux Beaux-Arts [a] été considéré comme inexistante." Letter from Dussaut to Jaujard, 13 June 1954. AN, AJ/37/486.

80 "fort préjudiciables à l'autorité de notre délégation." Letter from Jean Birlé to Delvincourt, Paris, 10 Nov. 1953. AN, AJ/37/486.

81 Letter from Birlé to Delvincourt, Paris, 10 Nov. 1953. AN, AJ/37/486.

82 "Pour éviter le retour à de semblables divergences de vues étalées au vu de l'étranger." Letter from Birlé to Delvincourt, Paris, 10 Nov. 1953. AN, AJ/37/486.

83 "beaucoup de musiciens considèrent que l'ancienne fréquence 435 Hz ne répond plus aux exigences artistiques actuelles." Letter from Birlé to Delvincourt, Paris, 10 Nov. 1953. AN, AJ/37/486. There seems to be some ground to Birlé's claim. In response to a German questionnaire, for example, an orchestra leader reported how "it has become increasingly customary for musicians to get used to the high musical pitch and to protest loudly whenever a normal musical pitch of *A* 435 Hz is required." ISO Technical Committee 43 Acoustics. German Proposal Concerning the standardization of Musical Pitch, 25 Oct. 1953. BBC, E2/437/2.

84 Letter from Delvincourt's secretary to Dussaut, 19 Nov. 1953. AN, AJ/37/486.

85 Project of letter from Dussaut to Birlé, 24 Nov. 1953. AN, AJ/37/486.

86 Copy of a letter from Cyril Ainsworth, Assistant Secretary and Technical Director of the American Standards Association, to Percy Good, Director of the BSI, n.d. Knudsen papers.

87 Letter from Robert W. Young to J. W. Mac Nair, 22 April 1949. Knudsen papers.

88 American Standards Association, Proposed American Standard Acoustical Terminology, Sponsor Acoustical Society of America in Cooperation with the Institute of Radio Engineers, Inc. (Washington, DC: American Standards Association, 1949). Quoted in "What Is the Pitch, Boys?," *Standardization* (April 1939): 101–102.

89 Harvey Fletcher, "Pitch, Loudness and Quality of Musical Tones." Bell Telephone System Technical Publication Monograph 13–1397 (reprint of a talk given before the American Physical Society and the American Association of Physics Teachers, in New York, 25 Jan. 1946). Lloyd read and commented on this publication now held in BBC Written Archives Center, R54/21/1.

90 As shown in chapter 1, Marloye, the maker who designed the first "diapason normal," collaborated with Savart for his experiments on hearing range and tone discrimination; and Despretz, one of the members of the 1858–1859 commission, teamed up with the maker to challenge Savart's findings. Work on musical pitch at MIT in the 1880s was similarly connected with studies of auditory perception, where it benefited from the knowledge, skills, and apparatus that Cross initially acquired to study hearing and speech (see chapter 3). As pointed out in chapter 4, finally, Lottermoser and Grützmacher contributed to the development of knowledge on auditory perception and integrated notions of hearing in their studies of musical instruments and musical pitch, although this did not find an immediate application in their efforts to standardize pitch. What is more, as scholars from Mersenne and Delezenne to Ellis and Lottermoser experienced, measuring frequencies was often a reciprocal operation during which one ended up observing their own ear. The presence of a tone variator indexed on the standard officially in use in Prussia at the time, *A* 435, in the collections of the physiology department at the Charité in Berlin shows how pitch standardization shaped the tools and practices of auditory perception. See https://soundandscience.de/image/tone-variator-no-1-detail-2-physiology-hearing-collection-charite.

91 Fletcher, "Pitch, Loudness and Quality of Musical Tones."

92 Mills, "Deafening"; Mills, "Decibel"; Sterne, *MP3*. On the history of hearing tests, see Viktoria Tkaczyk, Mara Mills, and Alexandra Hui (eds.), *Testing Hearing: The Making of Modern Aurality* (Oxford: Oxford University Press, 2020).

93 Hans-Joachim von Braunmühl and O. Schubert, "Ein neuer elektrischer Stimmtongeher für 440 Hz," *Akustische Zeitschrift* 6 (1941): 299–303. I use the BSI translation: BSI. Translation of photostatted article from *Akustische Zeitschrift* (*Acoustical Journal*), 1941. A New Electrical Tone Generator for 440 cps. BBC, R54/21/1.

94 BSI, *British Standards 880: Concert Pitch* (London: British Standards Institution, 1949), 14.

95 Memo from Pawley, 27 Feb. 1947. BBC, E2/437/2.

96 3M (B) 3777. Sub-Committee B/26/7—Concert Pitch of Technical Committee B/26—Acoustics. Unconfirmed Minutes of meeting, 31 Jan. 1949. BBC, R54/21/3. The two studies that Lloyd referred to are: E. G. Shower and R. Biddulph, "Differential Pitch Sensitivity of the Ear," *Journal of the Acoustical Society of America* 3 (1931): 275–287; and Seashore, *Psychology of Music*. In September 1946, Lloyd had sent Shower and Biddulph's article to Pawley. Letter from Wiggins to Pawley, 4 Sept. 1946. BBC, R54/21/1. On research on speech and hearing and Bell Laboratories, see Mills, "Deafening"; and Sterne, *MP3*. On the politics of Seashore's experiments, see Tkaczyk, "The Testing of a Hundred Listeners: Otto Abraham's Studies on 'Absolute Tone Consciousness,'" in *Testing Hearing*; and Martin Brody, "The Enabling Instrument: Milton Babbitt and the RCA Synthesizer," in *Contemporary Music Review* 39, no. 6, special issue "Opening the Doors of the Studio," ed. Jonathan Goldman, Fanny Gribenski, and João Romão (2020): 776–794.

97 Letter from Pawley to Lloyd, 20 June 1947. BBC, R54/21/2.

98 Pawley presumed that this was the reason why von Braunmühl and Schubert had designed tone generators rich in harmonics. As he explained to Alexander, and like several studies had demonstrated since the interwar period, "the variation of pitch with loudness appears to be less noticeable with complex tones than with pure notes." It was the two broadcasters' claim, however, that, according to the standards of studies on auditory perception, this choice was the result of empirical experiments. Memo from Pawley to Alexander, 30 May 1947. BBC, R54/21/2.

99 Letter from ABC's General Manager to the Director of the European Broadcasting Union, 30 March 1953. Quoted by Purtell, *Tuning the Antipodes*, 157.

100 Purtell, *Tuning the Antipodes*, 157.

101 BSI 3M (B) 748. Unconfirmed Minutes of meeting held at the Offices of the Institution [on 26 March 1947], 18 April 1947. BBC, R54/21/2.

102 ISO Technical Committee 43 Acoustics. German Proposal Concerning the standardization of Musical Pitch, 25 Oct. 1953. BBC, E2/437/2.

EPILOGUE

1 BSI. Technical Committee B/26. Acoustics. Report of proceedings at a conference on international agreement on concert pitch held under the auspices of the

British Standards Institution at Broadcasting House [on 20 May 1938]. Broadwood papers, 2185/JB/71/4a.

2 Francis Piggott, *The Music and Musical Instruments of Japan* (London: B. T. Batsford, 1893), 111.

3 The Moog synthesizer contained a vibrato function. So did the world's "first synthesizer," the electronic Sackbut, built in 1945 by Hugh Le Caine: https://ingeniumcanada.org/channel/innovation/electronic-sackbut-first-synthesizer.

4 The article was then shared with other members of BSI's pitch committee during their second meeting of 17 Oct. 1938, under the reference E (B) 9209. BSI. 2M (B) 2511. Unconfirmed minutes of second meeting held at the Offices of the Institution. Broadwood papers, 2185/JB/71/4a. Evelyn Broadwood's papers show evidence of the maker's further effort to document the history of pitch negotiation. The file dedicated to 1938–1939 conversations especially contains a fascicule on a 1885 presentation of tuning forks by the firm. *International inventions exhibition. Division—Music. 1885. John Broadwood and Sons.* On this article and Hipkins's contribution to the debates, see above, chapter 2.

5 The following British press reports on BSI and international efforts reflect the historical orientation of conversations: G. W. Kaye, "International Standard of Musical Pitch," *Nature* 142 (5 Nov. 1938): 820–821; by the same, "International Standard of Concert Pitch," *Nature* 143 (27 May 1939): 905–906; and Edward Gick Richardson, "The International Standard of Musical Pitch," *Journal of the Royal Society of Arts* 88 (20 Sept. 1940): 851–864. See also Llewelyn S. Lloyd, "International Standard Musical Pitch," *Journal of the Royal Society of Arts* 98 (1949): 74–89.

6 Bruce Haynes, *The End of Early Music: A Period Performer's History of Music* (Oxford: Oxford University Press, 2007).

7 Haynes, *The End of Early Music*, 44. For a critique of this demarcation between "baroque" and "modern" pitch, see Gérard Zwang, *Le Diapason* (Montpellier: Sauramp, 1998).

8 On the choice of 430 and 432 for nineteenth-century vocal works, see for example "Didone must be capable of everything. A conversation with Alessandro de Marchi on the challenge and necessity of resurrecting a Mercadante opera in its original sound." https://www.naxos.com/mainsite/blurbs_reviews.asp?item_code=NBD0095V&catNum=NBD0095V&filetype=About%20this%20Recording&language=English; and https://2019.theatrechampselysees.fr/en/season/staged-opera/la-traviata-en.

9 On these debates, see Ruth E. Rosenberg, "Perfect Pitch: 432 Hz Music and the Promise of Frequency," *Journal of Popular Music Studies* 33, no. 1 (March 2021): 137–154.

10 Ikeda's *A* project is presented here: http://www.ryojiikeda.com/project/A/. Ikeda's main source for his work is Ellis's 1880 article "On the History of Musical Pitch," as he explained during an interview with me on 13 Dec. 2017 in Los Angeles.

11 On this project, see Fanny Gribenski, "Sur le diapason: autour d'un cycle de création, entre son, science et lumière, de Nicolas Bernier, Dijon, Les Presses du réel, 2018, 176 pages," *Circuit: Musiques contemporaines* 31, no. 1 (2021): 89–91.

Index

Page numbers in italics refer to illustrations.

256 hertz, 11, 35
392 hertz, 190
405 hertz, 177
415 hertz, 15, 190
421 hertz, 190
430 hertz, 190
432 hertz: 435 hertz as rival to, 78–80; advocacy for, 5, 8; as standard *A*, 11, 75–76, 78–80, 174–80, 190, 191
433 hertz, 63
435 hertz: 432 hertz as rival to, 78–80; 440 hertz as rival to, 26, 119, 124, 126; as standard *A*, 3–4, 27, 50, 56, 65, 78, 80, 87, 89, 91, 96, 101, 106, 109, 118, 127, 129, 146. *See also* diapason normal; French pitch
438 hertz, 87
439 hertz, 90, 91, 158
440 hertz: 415 hertz as rival to, 190; 432 hertz as rival to, 174–80; 435 hertz as rival to, 26, 119, 124, 126; French opposition to, 174–80; political objections to, 5, 8, 119; as standard *A*, 1–6, 8, 20, 36, 45, 65, 90, 119–32, 134, 144, 156–66, 174–77, 182, 185, 190–92, 249n19
442 hertz, 185, 191
443 hertz, 191
444 hertz, 51, 191
445 hertz, 185
448 hertz, 47, 50, 51
449 hertz, 105
452.5 hertz, 91
455 hertz, 63, 67
460 hertz, 190
465 hertz, 190
500 hertz, 7, 31
512 hertz, 11, 66, 78
522 hertz, 65
528 hertz, 61, 65, 66–67
546 hertz, 65, 66

A (pitch): 392 hertz as standard for, 190; 405 hertz as standard for, 177; 415 hertz as standard for, 15, 190; 421 hertz as standard for, 190; 430 hertz as standard for, 190; 432 hertz as standard for, 5, 8, 75–76, 78–80, 174–80, 190, 191; 433 hertz as standard for, 63; 435 hertz as standard for, 3, 26, 27, 56, 65, 78, 80, 87, 89, 91, 96, 101, 106, 109, 118, 127, 129, 146; 439 hertz as standard for, 90, 91, 158; 440 hertz as standard for, 1, 4–6, 8, 20, 26, 36, 45, 65, 90, 119–32, 134, 144, 156–66, 174–77, 182, 185, 190–92, 249n19; 442 hertz as standard for, 185, 191; 443 hertz as standard for, 191; 444 hertz as standard for, 51, 191; 445 hertz as standard for, 185; 448 hertz as standard for, 47, 50, 51; 449 hertz as standard for, 105; 452.5 hertz as standard for, 91; 455 hertz as standard for, 63, 67; 460 hertz as standard for, 190; 465 hertz as standard for, 190; *C* as rival to, 43; as concert pitch, 1, 5, 9; controversies and variations in, 5, 8, 9, 11; material qualities of instruments as influence on use of, 11; pitch standardization linked to, 5, 11
Abbott, Frank D., *Musical Instruments at the World's Columbian Exposition*, *115*

ABC. *See* Australian Broadcasting Corporation
absolute music, 17
Academy of Ancient Music, London, 8
Academy of Fine Arts, France, 45, 177–79
Academy of Music, Boston, 101
Academy of Sciences, France, 30, 33, 178–80
Acoulon, Alfred, 174, 175–76
Acoustical Society of America (ASA), 140–42, 145
acousticians: in French standardization movement, 29–37, 52–53; interest in standardization as offshoot of other research of, 16–17; measurements taken by, 11–13; national associations of, 140–41. *See also* electroacousticians
acoustics: American interest and experiments in, 110–12, *111*, 129, 142; German interest and experiments in, 146–48; nineteenth-century French interest and experiments in, 16–17, 30–37; and physical embodiment of the standard, 52–53; precision in, 13, 15, 33, 35–36, 70, 84–86; standardization role of, 79–81, 142–43; Upham and, 102–3. *See also* psychoacoustics
aesthetic qualities: associated with pitch variation, 172–73, 185, 189; pitch standardization based on, 4, 6–10, 14, 17–19, 29, 40, 42, 138–39, 172–73
AFM. *See* American Federation of Musicians
AFNOR. *See* Association française de normalisation
Akustische Zeitschrift (journal), 142, 146
Alder, Ken, 52
Alexander, Francis William, 144, 157–58, 164, 183, 184, 242n54
All India Radio, 95
amateur musicians, 118, 128
American Federation of Musicians (AFM), 109, 118–20, 124, 126
American Guild of Organists, 127
American Physical Society, 141
American Standards Association, 142
Anderson, John, 127
Andrade, Edward, 164
anklung, 121
Arabic music, 95
Argus (newspaper), 164
ASA. *See* Acoustical Society of America
Association française de normalisation (AFNOR), 142–43, 174–77, 179–80, 183
AT&T, 141
Auber, Daniel-François-Esprit, 45
auditory perception, 15, 181–83, 255n90, 256n98
Australia, 163–64, 183–84
Australian Broadcasting Corporation (ABC), 163–64, 184
Austria: cultural aspirations of nineteenth-century, 67, 69–71, 79; France's relations with, 67–68; and the French pitch, 61, 68–74, 85–86; political conditions and practices in, 67–70; Prussia's relations with, 68; standardization efforts in, 67–74, 84–87, 222n110
Austrian Radio Company, 133
authenticity, 190

B (pitch), 7, 31, 44, 46
Bach, Johann Sebastian, 15, 100–101, 124, 153, 176, 177, 190
Barone, Alfonso, 169–74
Baroque music, 15, 138
BBC. *See* British Broadcasting Corporation
beat counting, 111, 122, 148
Beethoven, Ludwig van, 8, 100, 101, 106, 116, 124, 153
Beidler-Wagner, Franz Wilhelm, 138
Belgium, 74, 75, 78–79, 83
Bell, Alexander Graham, 110
Bell Laboratories, 170
Bell Telephone Company, 110, 176
Bénédit, Gustave, 40
Berlin Philharmoniker, 191
Berlioz, Hector, 7–8, 9, 14, 37–38, 45, 51, 52, 74; *Te Deum*, 38
Bernhart, José, 175–76
Bernier, Nicolas, *Frequencies*, 192
Besson, Gustave, 44
Biddulph, R., 182
Bijsterveld, Karin, 201n28
Birlé, Jean, 179–80
Bismarck, Otto von, 71, 79, 83
Blaikley, Arthur, 167, 173–74
Blaikley, David J., 92–93
Blaserna, Pietro, 78–79, 80, 82, 84
Bluthner & Co., 188
Bodin, François-Etienne, 44
Boito, Arrigo, 80
Bonaparte, Napoleon, 31, 60
Bonnet, Henri, 138
Bonnet, Joseph, 40–41
Boosey & Co., 92
Boston, Massachusetts, 99–108
Boston Music Hall Association, 102, 105–6
Boston Symphony Orchestra, 108, 191
Boult, Adrian, 15, 156, 182
Brahms, Johannes, 17
Braillard, Raymond, 136–37, 144–45
brass bands, 170
brass instruments: and brilliance, 9–10; pitch standardization for, 56; temperature's effect on, 126
Braunmühl, Hans-Joachim von, 143–44, 158, 170, 182–83, 256n98
brilliance (timbre), 9–10, 39, 90
Britain: *C* as foundational note in, 11, 61; and the diapason normal, 11; France's relations with, 60, 63; and the French pitch, 10, 61, 63–67;

imperial involvements of, 89–91; political conditions and practices in, 63–65; standardization efforts in, 63–67, 89–91, 163–69, 216n18; United States' relations with, 145
British Army, 90–91
British Broadcasting Corporation (BBC), 15, 133, 136, 138, 144, 151, 155–59, 163–69, 184
British Standards Institution (BSI), 136, 143–45, 155, 157, 162–69, 171, 173–77, 179–80, 182, 184, 190, 242n48
broadcasters. *See* radio broadcasters and sound engineers
Broadwood, Evelyn, 154, 189–90
Broadwood, Henry Fowler, 47
BSI. *See* British Standards Institution
Bureau des diapasons, Paris Conservatory, 53–54
Bureau of Standards (United States), 126, 128, 132, 142, 157

C (pitch): 256 hertz as standard for, 11, 35; 512 hertz as standard for, 11, 35, 66, 78; 522 hertz as standard for, 65; 528 hertz as standard for, 61, 65, 66–67; 546 hertz as standard for, 65, 66; as foundational note in Britain and United States, 11, 61, 65, 66–67; mathematical and scientific arguments for pitch standardization on, 35, 43, 66, 78
Cabarat, Robert, 175
Cage, John, 22
canonization, and pitch standardization, 9, 40, 42, 50, 138, 152–53, 155, 171
"Can We Have an 'A'?" (radio program), 166–69
Carrillo, Julián, 171
Case Western University, 83, 128–29
Castil-Blaze (François-Henri-Joseph Blaze), 9
cathedral chimes, 121
Cavaillé-Coll, Aristide, 38, 51, 53, 91, 205n14
C. Bruno & Sons, 127
certification, of conformity to standards, 173–74
Chartier, Richard, *Transparency (Performance)*, 192
Chester, Harry, 63–64
Chicago Symphony Orchestra, 191
Chickering & Sons, 127
China, 88
Chladni, Ernst Friedrich, 32, 36, 45, 52, 204n7; *Die Akustik*, 30
CIA. *See* Comité International Acoustique
clarinet, 126
classical music: cultural value associated with, 101–2, 131, 138, 152, 188; pitches associated with, 106, 112, 116, 120, 126, 131–32, 137–38, 153–54; popular music as rival to, 120, 126; radio broadcast of, 137–38
climate. *See* temperature and climate
clock fork, 84, 122, 130
Coccia, Carlo, 47
colonialism. *See* imperialism/colonialism
color. *See* timbre
Comité International Acoustique (CIA), 133–34
Comité national d'Acoustique, 142–43
comparative musicology, 96
complex tones, 183–84, 256n98
composing, in response to pitch standardization, 191–92
Concentus Musicus, Vienna, 190
concert pitch: aesthetic and cultural concerns underlying standardization of, 7–10, 187; controversies and variations in, 5–6, 155, 167–68, 183–86, 190–92; electroacoustic measurement of, *150*; implementation of, 164–69, 173–74; political and social significance of, 18, 20; radio's role in the setting of, 151–59, 164–69; universalist aims vs. local conditions and practices, 187–89. *See also* 440 hertz
Concerts Musard, Paris, 44
conferences on pitch standardization: Belgium (1884), 79; London (1939), 1–2, 134, 136–39, 141, 143–46, 154–55, *154*, 156, 160, 161, 164, 169, 175, 184, 190; London (1953), 179–80, 184, 186; Milan (1881), 78–79; Vienna (1885), 36, 61, 74–75, 79–83, 85, 89, 142, 146, 156
Conn (firm), 127–28
conspiracy theories, 5, 8, 159–60
Couesnon (firm), 176
Couvreur, Jacques, 175–76
Crimean War (1853–1856), 60
Cross, Charles R., 17, 110–13, 149, 230n58, 255n90
Cushman, Charlotte, 99

Deagan, John C., 120–29, 235n118; *Catalog R*, 124, *125*
Dea-gan-ometer, 122, *123*
Debain, Alexandre, 44
decibel, 141
Delezenne, Charles, 12, 45, 52
Delvincourt, Claude, 176, 178–80
Desains, Paul, 33
Despretz, César-Mansuète, 30, 32, 45–47, 50–51, 255n90
De Traz, Robert, 133
Deutscher Akustische Ausschuß, 140
Deutschmann, Charles, 127–28
diapason normal: British use of, 11, 89–90; coining of the term, 201n22; creation of, 4, 7, 8, 43–57; criticisms of, 62; failures of enforcement of, 53–54; geographical spread of, 4; institutionalization of, 52; international negotiations concerning, 74–83; meanings of "diapason," 10; temperature as factor in, 11, 13; vocal basis of, 4. *See also* 435 hertz; French pitch
Dilke, Charles Wentworth, 63–64
Disney, 121
Donizetti, Gaetano, 40
Doucet, Camille, 45

Duff-Challen, Margaret, 167–68
Dumas, Jean-Baptiste, 45
Dumesnil, René, 177
Dussaut, Robert, 174–80, 186, 190
Dutch Foundation, 141
Duval, Charles, 174, 175–76
Dwight, John Sullivan, 102, 105–6, 108
Dwight's Journal of Music, 102, 105–6, 108

early music, 5, 8, 186, 190
economics: of instrument making, 32; of organ building, 3; of pitch standardization, 13–14, 55–56, 131
education, 60, 75
Einstein, Albert, 129
electro-acousticians: international networks of, 142–43; pitch measurement by, 25, 149–55, *150*; standardization role of, 4, 7, 134–35, 137, 159. *See also* acousticians
Ellis, Alexander J., 7, 24, 76, 96, 111, 112, 116, 124, 149; "On the History of Musical Pitch," 76, *77*; "On the Musical Scales of Various Nations," 96
enforcement of standards. *See* implementation/enforcement of standards
Engineering Standards Committee (Britain), 142
Engineering Standards Committee (United States), 142
entertainment industry, 120–21
environmental acoustics, 22
equal temperament, 11–12, 19–20, 87, 88, 95, 124, 148, 171
Esclangon, Ernest, 180
ethnomusicology, 24
European Economic Community, Committee of Ministers, 8

Farrenc, Aristide, 44
Fétis, François-Joseph, 39–40
Figuier, Louis, 34
Finkelstein, Gabriel, 31
Flaubert, Gustave, 91
Fletcher, Alice Cunningham, 96
Fletcher, Harvey, 140, 181–82
Fokker, Adriaan Daniël, 141
Foucault, Michel, 189
France: Austria's relations with, 67–68; Britain's relations with, 60, 63; and international relations, 60; and the pitch quarrel, 174–80, 186; political conditions and practices in, 45, 55–56, 60–61, 69; scientific and cultural aspirations and status of, 30–32, 45–46, 57, 60–61, 63, 75, 174–75; standardization efforts in, 3–4, 7, 10, 27–57, 161, 174–80, 187; vocal music's significance in, 39–40, 49
Franche, Charles Louis, 44
Franco-Austrian War (1859), 60, 67–68
Franco-Prussian War (1870), 60
Franz Joseph I, 67–68
French Acoustical Committee, 152
French pitch: Austria and, 61, 68–74, 85–86; British resistance to, 10, 90; controversies and variations, 51–57, 62–63; double origin of, 30–43; institutionalization of, 43–51; international spread of, 4, 59–83; in United States, 101, 105–19. *See also* 435 hertz; diapason normal
frequency, vs. pitch, 14, 181–82, 192
Fuller, Levi K., 109–10, 112–13, 115–17, 230n57

G (pitch), 35
geography: and pitch standardization, 10–11; and spread of diapason normal, 4; temperature as consideration in, 13
George V, 91
Germania Orchestra (Boston), 101
German pitch. *See* Stuttgart pitch
Germany: Austrian–Prussian conflict in, 68–70, 73; Austria's aspirations for political and cultural influence in, 69–74; *A* as foundational note in, 11; and the French pitch, 71–74; German Confederation, 61, 67–68, 70–72, 217n40; Nazi, 5, 137, 139, 146–47, 159–60, 177; political conditions and practices in, 71–74, 146–47; standardization efforts in, 71–74, 79, 83, 185–86; United States' relations with, 119
Gevaert, François-Auguste, 39
Gewandhaus, Vienna, 191
Giacomelli, Adolphe, 44
globalization: and controversies over standardization, 20–21; of music, 21–22; and pitch standardization, 3, 4, 11, 22, 59–61, 69–71, 159. *See also* imperialism/colonialism; international relations
Gluck, Christoph Willibald, 40, 50
Goebbels, Joseph, 5
Good, Percy, 162
Graves, Cecil, 136, 138–40, 156
Great Organ, Music Hall, Boston, 99–108, *100*, 110
Greece, 75
Greenwich pips, 157
Grützmacher, Martin, 135, 143, 145–48, 151, 177, 255n90
Gulbransen, 128

Habeneck, François, 41–42, 50
Halévy, Fromental, 5, 45
Handel, George Frideric, 103, 116, 190
Handel and Haydn Society, 101, 105–6
Hanslick, Eduard, 17–18, 59, 68–69, 73, 88
harmoniums, 92, 94–95
Harnoncourt, Nikolaus, 190
Harvard Musical Association, 101–2, 106
Harvard University, 101–2, 106
Haunted House, The (cartoon), 121
Hawkes, William Henry, 90–91, *91*
Haydn, Joseph, 8, 190

Hayes, L. W., 133–35, 143–45, 156, 159
Haynes, Bruce, 190
Helmholtz, Hermann von, 76, 124
Herschel, John, 65–66, 78
hertz, defined, 4
Hill, William, 104
Hipkins, Alfred, 90, 116, 124, 126, 190
Hitler, Adolf, 139
Horowitz, Joseph, 105
Houdek & Hervert, 86
Hunt, Bruce, 52

IBU. *See* International Broadcasting Union
Ikeda, Ryoji, 192
imperialism/colonialism: American, 119; competing interests affecting pitch standardization in, 89; keyboard instruments and, 91–92; music's role in, 88–97; and pitch standardization, 4, 13, 22, 88–89, 92–97, 189; and the soundscape, 23; temperature as consideration arising from, 13, 62
implementation/enforcement of standards: in Austria, 85–86; in Britain, 63–64, 163–69; challenges of, 19, 33, 36, 51–56, 60, 63–64, 71–73, 75–76, 87–88, 130–31, 162–73, 183, 189; in France, 51–56, 60, 81; in Germany, 36, 71–73, 182; international, 75–76, 81–88, 144, 159, 162–65, 169, 175–76; physical constraints on, 4, 12; radio as means of, 145–60, 163–69; temperature and climate as factors in, 81–82, 155; in United States, 116–18, 126, 129–31
Indian music, 94–95
Inghelbrecht, Désiré-Émile, 176
Institut International du Son, 175, 176
instrument makers: criticisms of, 9–10, 32; industrialization undergone by, 14; influence of science on, 120–21; pitches used by, 113; and pitch standardization, 14, 43–47, 55, 60, 117–18, 122, 127–31; precision sought by, 60; regulation of, 130
instruments: pitch standardization for, 3, 28, 47, 49, 171–72; temperature's effect on, 13, 82, 92–93, *94*, 126, 155, 167–68; timbre of, 39. *See also* brass instruments; instrument makers; materiality of instruments; organs; percussion instruments; string instruments; wind instruments
interdisciplinary studies, 17–18
International Broadcasting Union (IBU), 136–38, 143–45, 151, 153, 155, 158, 160, 240n15; Checking Center, 151, 153
International Electrical Congress, 52
International Electrotechnical Commission, 134
International Musical Society, 135
International Organization for Standardization (ISO), 20, 134, 161–64, 169, 173, 179, 184–86, 188
international relations: France and, 60; music's role in, 24–25; pitch standardization and, 69–70, 74–88, 135–40, 143–45; tensions in, 63, 75, 135–40. *See also* globalization; political and social issues
International Standards Association (ISA), 134, 143–44, 151, 160, 164; Subcommittee 43/3[B], 135–37, 143–46, 155, 157, 240n9
ISO. *See* International Organization for Standardization
ISO 16, 4, 20, 180
Italian Opera, London, 65
Italy: *A* as foundational note in, 11; standardization efforts in, 75, 78–80, 83, 84–85, *85*, 170–71

Jackson, Myles, 16, 80–81, 201n28
James, Henry, 99
Japan, 188
Jaujard, Jacques, 179–80
jazz, 10, 87, 119, 131, 152–53, 177, 200n15
Journal des débats (newspaper), 38
Journal of the Acoustical Society of America, 142
just intonation, 11, 200n18

keyboard instruments: climatic effects on, 92; in colonial settings, 94–96. *See also* organs; pianos
Kneller Hall Pitch, 90
Knudsen, Vern Oliver, 140, 145, 157
Koenig, Rudolph, 53, 78, 84–85, 110, 116–17, 122, 130, 192, 221n102
Köhler, Augustus Charles, 90
Kotykiewicz, Teofil, 87
Kreidl, Alois, 86
kymograph, 149

Labrousse, André, 152–53
La Fage, Adrien de, 42, 44–45, 51, 53
La France musicale (journal), 27
La Musica (journal), 51
Lanckoronski, Karol, 70
League of Nations, 135, 138
Le Maistre, Charles, 136, 145
Lerebours and Secrétan (instrument makers), 36, 53, 73, 221n102
Lissajous, Jules-Antoine, 7, 16–17, 30–37, 43–47, 50–51, 53, 83, 111, 138, 149, 187, 204n6, 208n43; *Mémoire sur l'étude optique des mouvements vibratoires*, *34*
Liszt, Franz, 17
Lloyd, Llewelyn S., 139, 161, 165, 167, 170–71, 173, 181, 182
Lockyer, W., 65
London Philharmonic, 47
Lottermoser, Werner, 146–53, 170–71, 243n61, 255n90
loudness, 183
Louisiana Purchase Exposition (St. Louis, 1904), 118

Louis XIV, 8, 32, 42, 50
Loulié, Étienne, 202n35
Löwenherz, Leopold, 84
Ludwig, Carl Friedrich Wilhelm, 149
Lunn, Henry, 67
Lvov, Alexei, 74–75
Lyon & Healy, 127

makers. *See* instrument makers
Marchal de Calvi, Charles-Jacob, 27–28, 32
marimbas, 120, 121
Marloye, Joseph, 30, 32, 36, 52, 201n22, 255n90
Mason, W., 65
Massachusetts Institute of Technology (MIT), 110, 111, *111*, 255n90
material embodiment of standards. *See* physical embodiment of standards
materiality of instruments: pitch standardization constrained by, 3, 11, 12–13; temperature's influence on, 13; timbre influenced by, 148
mathematical basis for pitch standardization, 11, 30, 35–36, 65–66, 78, 178. *See also* science and technology
meantone temperaments, 11, 200n18
Meerens, Charles, 11–12, 76, 78–79
Melba, Nellie, 89, 164
Mellinet, Émile, 45, 56–57
Mendelssohn, Felix, 106
Mengelberg, Willem, 153
Mersenne, Marin, 3, 21, 34–35, 59
metric system, 7, 21, 31, 44, 52–53, 78
metronome, 202n35
Metternich, Richard von, 68
Mexico, 164
Meyer, Erwin, 141, 147
military bands: economic factors in pitch standardization for, 72, 90; musicians in, 90; pitches used by, 3, 38, 44, 45, 56–57; prices of instruments for, *91*; as vehicles of imperialism, 92
Miller, Dayton Clarence, 128–30, 141
Miller, Henry J., 110–13
Mills, Mara, 201n28
Ministry of Education, France, 177–79
MIT. *See* Massachusetts Institute of Technology
Møller, Mathias Peter, 127
Monnais, Édouard, 45
Montal, Claude, 44, 50
Morgan, Augustus de, 65
Mozart, Wolfgang Amadeus, 8, 116, 190; *Magic Flute*, 8
Murphy, Craig, 142
Murphy, O. J., 170, 173
Musée de la Musique, 53
music: absolute, 17; considerations in standardization of, 2, 4, 6–7, 28, 51, 62, 161, 162–63; globalization of, 21–22; historical approaches to, 8–9, 12; imperial role of, 88–97; peace goals pursued through, 24–25, 135–39. *See also* canonization; musicology; Western music
musical education, 60, 75
Musical Fund Society Orchestra (Boston), 101
musical instruments. *See* instruments
Music Hall, Boston, 99–108, *100*, *104*
musicians: international travel of, 60, 137; perception and behavior of, 181–85; pitch adjustments made by, 60, 168; and the pitch quarrel, 175–76, 178–79, 254n73; standardization role of, 29, 37–43, 50–51, 55, 164–65, 168, 173–74, 178–79, 181–82
Music Industries Chamber of Commerce (United States), 127–29
musicology: and alleged ascent of pitch, 7; comparative, 96; Ellis's influence on, 24; material turn in, 20; scholarly interest in pitch standardization, 15–18; sound studies in, 23
Music Supervisors' National Conference, 127
Music Teachers' National Association, 127
Music Trade Review (magazine), 118, 119, 128
Mussolini, Benito, 139, 170

Napoleon III, 25, 28, 31, 45, 60–62, 67–68
National Association of Organists, 127
National Association of Piano Tuners, 127
National League of Musicians, 109
National Music Teachers' Association, 109
National Piano Technicians Association, 128
Native American music, 96
nature: musical technologies used in scientific study of, 37; standards justified by appeal to, 2, 5, 12, 21
Nature (magazine), 112
New Age, 159
New England Conservatory, 108
New Philharmonic Pitch, 89, 168
New York Herald Tribune (newspaper), 161–62, 184
noise, 141
Nordwestdeutscher Rundfunk, 170
"Normal Diapason!" concert (Boston, 1869), 106, *107*, 108

oboe, 1, 25, 126, 155, 183, 187, 190–91
Old Philharmonic Pitch, 47, 89, 91
orchestras: broadcasting of, 137–38; new conceptions of, 9; sound of, 39–40, 137–38; tuning of, 1, 5. *See also* concert pitch
organs: in colonial settings, 91–92; economics of, 3; Great Organ, Music Hall, Boston, 99–108, *100*, 110; makers of, 109, 127, 129; pitch standardization for, 11, 109, 127, 171; temperature's effect on, 82; as touchstone for tuning, 3; tuning practices for, 37

Paganini, Niccolò, 9
Paine, John Knowles, 106

Paris: variations of pitch found in, 3, 37–38, 41; vocal music's significance in, 39–40, 49
Paris Conservatory, 3, 19, 40–42, 44, 53, 60, 75, 81, 129, 174–76, 178
Paris Opera, 22, 27, 32, 40–42, 47, 50, 51, 52, 54–55
Paris Opéra-Comique, 38
Paris Théâtre-Italien, 38, 41
Patti, Adelina, 67
Pawley, Edward, 157–58, 162, 163, 165–69, 183, 249n19, 256n98
peace, international standardization linked to, 24–25, 135–40
perception. *See* auditory perception
percussion instruments, 120–22
perfect pitch, 15, 168
performance practice studies, 24
Philharmonic Academy, Turin, 47
Philharmonic Society (Britain), 63, 90
Philips Corporation, 144, 151
phonodeik, 129–30
Phonogramm-Archiv, 96
physical embodiment of standards, 10, 14, 52–53, 83
Physikalisch-Technische Reichsanstalt (PTR), Berlin, 83–84, 129, 146–48, 151, 164
Physikalisch-Technische Reichsanstalt (PTR), Braunschweig, 170–71
Pianomania (documentary), 191
Piano Manufacturers' Association, 109, 113, 127, 237n135
pianos: in colonial settings, 92, 95; makers of, 109–17, 127–29; pitch standardization for, 109–17, 127, 171; tuners' effect on aural aesthetic of, 191
Piggott, Francis Taylor, 188
pitch: in colonial locations, 4, 13, 22, 88–89, 92–97; color linked to, 9; diverse concepts of, 2–3, 29; escalation of, 7–10, 12, 14, 32, 38–41, 47, 49–50, 62, 65, 67, 69, 70, 75, 106, 152, 161, 176, 209n51; frequency distinguished from, 14, 181–82, 192; geography concerns involving, 10–11; historical/comparative approach to, 7–10, 12, 40–42, 76, 96, 113, *114*, 115–16, 124, *125*, 139, 170, 190, 192; international negotiations on, 74–83; issues at stake in standardization of, 2, 6–7; linguistic factors affecting, 10–11; "live" evaluation of, 149–50; loudness as factor in, 183; measurements of, 10–11, 148–55; post–World War II research on, 170–74; subjective character of, 4, 14, 181–82; temperature's effect on, 11, 13, 81–82, 90; variations in, 12, 13, 15, 37–38, 51, 56, 67, 87, 112, 138, 171–73, 189; visual measurements of, 151–52; voice and, 3, 32. *See also* concert pitch; French pitch; perfect pitch; standards and standardization; tuning
pitch quarrel, 174–80, 186
Platen-Hallermund, Count von, 72
Pol, Balthasar van der, 144, 151–53, 156, 158, 170–71
Pole, William, 65
political and social issues: in Austrian standardization efforts, 67–70; in British standardization efforts, 63–65; conspiracy theories, 5, 8, 159–60; in French standardization efforts, 45, 55–56, 60–61, 69, 174–80; in German standardization efforts, 71–74; in international standardization efforts, 7, 63, 75, 83–88, 135–40; in Nazi Germany, 146–47; peace goals of international standardization, 24–25, 135–40; the pitch quarrel, 174–80; in pitch standardization, 2, 5, 8, 13–14, 18, 20–21, 119, 159–60; technical standardization efforts linked to, 135–40. *See also* international relations
popular music, 5, 13, 88, 99, 119–21, 126–27, 131, 153, 177, 191
Pouillet, Claude, 34
Praetorius, Michael, 3
precision: acoustic, 13, 15, 33, 35–36, 70, 84–86; aesthetic undesirability of, 172–73; in broadcast of 440 hertz signal, 15, 182; as goal of standardization, 2; in instrument making, 60; in musical pedagogy, 60; in popular music, 191; in tuning fork manufacture, 60, 84, 86, 182; in Western music, 19–20, 188
Prony, Gaspard de, 36
Prussia, 13, 60, 68, 70, 71, 73
psychoacoustics, 14, 169, 181–83
PTR. *See* Physikalisch-Technische Reichsanstalt
public, awareness and adoption of standards by, 14–15, 164–69, 249n19
Pythagorean tuning, 12, 78, 178, 200n18

Rabaud, Henri, 175
racism, 177
radio broadcasters and sound engineers: and concert pitch, 151–59; friendships among, 145; international community of listeners created by, 136, 138; limitations constraining, 153; measurement capabilities of, 13; networks of, 140–45; standardization role of, 4, 7, 132, 135, 137–39, 145–60, 163–64
Radiodiffusion, 175, 176
Rambert, Maurice, 156, 158
Rechberg und Rothenlöwen, Bernhard von, 70–71
reed organs, 92
Reichs-Rundfunk-Gesellschaft (Reich Radio Society), 143, 151, 158, 182
Rhohrer, Jérémie, 190
Richard, P.-J., 176
Robinson, Stanford, 167–68
Rossini, Gioachino, 42
Royal Academy of Music (Britain), 89
Royal Chapel, Versailles, 8, 41–42

Royal Military School of Music (Britain), 90
Royal Society of Arts. *See* Society of Arts (Britain)
Russia, 74–75
Russian State University, 84

Saint-Eustache, Paris, 38
Saldman, Charles, 65
Sandoz, Albert, 53, 81
Sauveur, Joseph, 11, 35, 178
Sax, Adolphe, 57
Schaeffer, Pierre, *Diapason Concertino*, 192
Schafer, R. Murray, *The Soundscape*, 22–23
Schaffer, Simon, 16, 52
Scheibler, Johann Heinrich, 36, 52, 111, 122, 207n36
Schmerling, Anton von, 70, 217n47
Schubert, O., 182–83, 256n98
Schumann, Robert and Clara, 74
Schwaiger, Gustav, 133–34
science and technology: in Britain, 65; in France, 29, 50–51; and instrument making, 120–21; music in relation to, 18–19; popularizers of, 208n43; relevance of, to standardization history, 15–20; scholarship on standards in, 16; standardization role of, 29, 50–51, 65, 81, 111, 128, 146–52, 173, 184–85. *See also* mathematical basis for pitch standardization
Seashore, Carl, 183
SEIN. *See* Société d'Encouragement pour l'Industrie Nationale
simple tones, 183–84
Skinner, Ernest, 127
Snell, George, 102–3
Société d'Encouragement pour l'Industrie Nationale (SEIN), 31–32, 43, 45
Société des facteurs de piano, 43
Société musicale de Caen, 56
Société syndicale des fabricants de pianos et autres instruments de musique, 45
Society of Arts (Britain), 11, 61, 63–64, 73, 89–90, 105
sound engineers. *See* radio broadcasters and sound engineers
Sound Foundation (Netherlands), 140
soundscape, 22–23
sound studies, 23
South African Broadcasting Corporation, 155
Spain, 74, 75, 83
sphygmosphone, 103
sruti boxes, 95
Staffa, Giuseppe, 51
Stainer, John, 89–90
standards agencies, 142
standards and standardization: brief history of, 2–4, 35; China and, 88; colonialism and, 4, 13, 22; competing interests affecting, 2, 4, 6–7, 14, 29–30, 37, 43, 66, 89, 117; composing in response to, 191–92; conspiracy theories about, 5, 8, 159–60; controversies and variations in, 5–6, 14, 51–52, 55–56, 67, 70–74, 81, 85, 87–89 (*see also* limitations of); economics of, 13–14, 55–56, 131; France's role in, 3–4, 7, 10, 27–57, 161; geographic concerns in, 10–11; globalization and, 3, 4, 22, 59–61, 69–71, 159; historical approach to, 7–10, 12; impact of, 15–16; of instruments' pitch, 3, 28; internationalization of, 4, 126, 130–32, 134–40, 143–45, 159, 161–62, 175–76, 179, 188; limitations of, 14–15, 57, 85–86, 88, 93, 169–74, 184–85, 187, 189 (*see also* controversies and variations in); material qualities of instruments as influence on, 11, 12; mathematical basis for, 11, 30, 35–36, 65–66, 78, 178; metric system as model for, 7, 21, 31, 44, 52–53, 78; music's unique considerations regarding, 2, 4, 6–7, 28, 51, 62, 161, 162–63; nature cited as authority for, 2, 5, 12, 21; in optics, 35; peace linked to, 24–25, 135–40; physical embodiment of, 10, 14, 52–53, 83; post–World War II, 161–86; public attitudes toward, 14–15, 164–69; purpose of, 2; scholarship on, 15–17; subjective character of, 4, 14; theoretical basis of, 12, 29–30, 51, 62, 65–66, 78, 80, 159; unintended effects of, 21, 24, 63, 73–74, 81, 192; universalist aims vs. local conditions and practices, 7, 12–13, 21, 24, 76, 97, 138, 192; various measuring conventions as influence on, 10–11; of vocal pitch, 4, 7, 27–28, 49–50. *See also* conferences on pitch standardization; diapason normal; French pitch; implementation/enforcement of standards; pitch
Stefan, Josef, 80, 83–84
Steflitschek, Franz, 86
Steinway, William, 109, 111, 113
Steinway & Sons, 127
Sterne, Jonathan, 201n28
Stewart, George W., 118
string instruments: pitch standardization for, 43, 54–55, 75, 170, 214n117; sound produced by, 75; wind instruments compared to, 168
Stuttgart pitch, 36, 45, 51, 65, 119
Sudre, René, 177
Sweden, 74
Symphony Hall, Boston, 99

Tagore, Rabindranath, 95
Teatro del Fondo, Naples, 51
technology. *See* science and technology
temperature and climate: measurement of, 82; musical instruments affected by, 13, 82, 92–93, *94*, 126, 155, 167–68; pitch affected by, 11, 13, 81–82, 90, 155; tuning affected by, 10–11, 62; tuning forks affected by, 11, 13, 52, 84–85, 112, 155
tempered scale, 11, 19
tempering. *See* equal temperament; meantone temperaments

tempo, standardization of, 19
Thomas, Ambroise, 45
Thomas, Theodore, 111
Thompson, Emily, 99, 201n28
Tiby, Ottavio, 169–74, 250n28
timbre (tone color), 9, 39, 65, 139, 148, 172, 182, 183. *See also* brilliance
time, 11. *See also* tempo
Tobin, J. Raymond, 168
tone color. *See* timbre
tone generators, 159, 163–64, 182–84, 256n98
tonometer, 122, 192
Tons de chapelle and *de chambre*, 37
Trade Review (magazine), 128
Treaty of Versailles (1919), 18
Triébert, Charles Louis, 44
tuning: aesthetic significance of, 191; American practices of, 112; controversies and variations in, 5–6, 11–12, 171–72; economics of, 3; equal temperament used in, 11; historical/comparative approach to, 7–10, 42, 46–49; increased brilliance as goal of higher, 9; local practices of, 3; loudness as factor in, 183; of orchestra instruments, 1, 5; temperature's effect on, 10–11, 62; various measurement considerations as influence on, 10–11. *See also* concert pitch; pitch
tuning devices: blamed for pitch escalation, 32–33; Deagan's invention of, 122; demonstrations involving, 16–17, 30, 32–33, *34*, 35, 113, 115–16; for the diapason, 10; gold construction of first French standard, 53, *54*; history of, 6–7; manufacture of, 53, 73, 83–86, *85*, 116–17, 231n67; nonmusical uses of, 37; precision of, 130; standardization role of, 5, 6, 13, 25, 35, 36, 42, 53, 83; temperature's influence on, 11, 13, 52, 112; used in comparative study of tuning practices, 46, *47*, *48*
Turnage, Mark-Anthony, *About Time*, 192
Tyndall, John, 208n43

United Nations Standards Committee, 161
United States: Britain's relations with, 145; *C* as foundational note in, 11; cultural aspirations of, 119; European culture in relation to that of, 99–101, 118, 177; and the French pitch, 101, 105–19; Germany's relations with, 119; imperialism of, 119; international standardization role of, 4, 126, 130–32, 160, 177; standardization efforts in, 105–32
University La Sapienza, Rome, 83, 129
University of Rennes, 192
University of Vienna, 83–84, 129
Upham, Jabez Baxter, 102–6, 227n17

Valentine and Carr (tuning-fork makers), 116–17
valve oscillator, 148
Varèse, Edgar, *Tuning Up*, 192
Verdi, Giuseppe, 78, 80, 190
vibraphone, 236n128
Victoria, Queen, 90
Viennese Männergesangverein, 87
Viennese Philharmonic Orchestra, 86
Viennese Schubertbund, 87
voice: pitch linked to physiological capacity of, 3, 27–28, 32, 39, 40, 65; pitch standardization for, 4, 7, 27–28, 49–50, 188
Vrijzinnige Partij (Netherlands), 5

Wagner, Fritz, 86
Wagner, Richard, 17, 116
Wagner Festival, 67
Walcker, Eberhard Friedrich, 91, 99, 104, 108
Wang Mang, 88
Weber, Max, 189
Weber, Wilhelm, 35–36, 52–53
Western music: colonies' responses to, 93–96; conceptions of, 24; desire for precision in, 19–20, 188; economic influences on, 131; pitch standardization applicable to, 1, 5, 21, 152, 188. *See also* canonization; imperialism/colonialism
Westminster Abbey, London, 103
Wheatstone, Charles, 34–35, 65
Whewell, William, 65
White, William Braid, 1, 19, 119, 127–30
Wiggins, Ethel, 163
Wilhite, Oscar Milton, 118
Willis, Henry, 91
wind instruments: and brilliance, 9; in colonial settings, 92–93; manufacture of, 91; pitch standardization for, 52, 55, 90–91, 171; string instruments compared to, 168; temperature's effect on, 82, 90, 93, 155. *See also* brass instruments
Wolfe, John, 167
World's Columbian Exposition (Chicago, 1893), 115, *115*
World's Fair (Paris, 1867), 56
World War I, 25, 135, 137
World War II, 4, 134–35, 137, 139–40, 144, 159–60, 169
Wunderer, Alexander, 87–88

xylophone, 120, 121

Yates, JoAnne, 142
Young, Robert W., 181
Young, Thomas, 34

Zarlino, Gioseffo, 178, 200n18
Zeller, Carl, 80
Ziegler, Henry, 127
Zwikker, Cornelis, 141